Combustion in Piston Engines
Technology, Evolution, Diagnosis and Control

Antoni K. Oppenheim

Springer

Berlin
Heidelberg
New York
Hong Kong
London
Milan
Paris
Tokyo

Engineering

springeronline.com

A. K. Oppenheim

Combustion in Piston Engines

Technology, Evolution, Diagnosis and Control

With 96 Figures and 12 Tables

 Springer

Antoni K. Oppenheim
Professor of Engineering
Department of Mechanical Engineering
University of California
5112 Etcheverry Hall
Berkeley, CA 94720-1740
USA

E-Mail: ako@me.berkeley.edu

On the cover
Photograph of a distributed ("homogeneous") turbulent combustion field,
created away from the walls of its cylindrical enclosure by an opposed
triple-stream FJI&I (Flame Jet Injection and Ignition) system (vid. Fig. 2.15)

ISBN 3-540-20104-1 Springer-Verlag Berlin Heidelberg New York

Cataloging-in-Publication Data applied for
Bibliographic information published by Die Deutsche Bibliothek.
Die Deutsche Bibliothek lists this publication in the Deutsche Nationalbibliografie;
detailed bibliographic data is available in the Internet at <http://dnb.ddb.de>.

Springer-Verlag Berlin Heidelberg New York
Springer-Verlag is a part of Springer Science+Business Media

springeronline.com

© Springer-Verlag Berlin Heidelberg 2004
Printed in Germany

Typesetting: Data conversion by author
Final processing by PTP-Berlin Protago-TeX-Production GmbH, Berlin
Cover-design: de´blik, Berlin
Printed on acid-free paper   02 / 3020

Preface

Upon an exponential growth of life on earth, supported by a century of oil-based energy economy, following a century of its coal-based phase, the world we live in is facing a point of inflection[1] at the threshold of an exponential decay. It is, in fact, the quest for counteracting its dreaded consequences that are bound to follow, which half a century ago provided the stimulus for launching the space program.

Coal and oil are made out of carbon and hydrogen atoms. Hydrocarbon molecules consisting of these atoms are the essential ingredients of life by providing energy in a remarkably compact from. This property is due to the fact that hydrocarbons, by themselves, are not highly energetic materials. The energy is derived from them by chemical reaction with oxygen supplied by the surrounding air. This reaction provides heat and is referred to for that reason as exothermic. The mass of oxidizer it requires is, in fact, ~ 3.5 (if it is provided by oxygen alone) or ~ 15 (if it is supplied by air) times larger than that of the hydrocarbon. For vehicular transport, where the fuel has to be carried on board, this feature is of particular significance. The simplest way to get a sufficient amount of power (rate of energy) for this purpose is by carrying out this reaction by combustion – an oxidation of fuel that takes place when the temperature is elevated above a threshold level of about 1000 K (700 °C) at atmospheric pressure. This is then the raison d'être for internal combustion engines – the prime automotive powerplant today.

Concomitantly, with the initial conquest of space by voyage to the moon – a triumph of technological advances attained half a century ago – the industrial world became concerned with the escalating generation of pollutants by combustion and the inevitably dwindling natural resources of energy. As a consequence, the automotive industry found itself confronted with a significant amount of societal pressures and governmental regulations. This brought about a state of flux and confusion. Its most prominent

---

[1] Oppenheim AK (1992) Life on earth at the point of inflection. Proc. Inst. Mech. Eng. London, C448/076, pp. 215–220
www.rognerud.com/pollution/html/earth.html

manifestation was a general disenchantment with internal combustion engines and the frantic quest, stimulated thereby, for alternatives, such as alternative fuels, alternative powerplants; alternative energy resources, and anything except for what we have today. Interestingly enough, the more vague is the implementation of these alternatives, the more are they in vague and, hence, the more are they financially supported.

An overall view upon the evolution of energy resources and technology for road vehicles is displayed by Table 1.

Table 1. Energy resources and technology for road vehicles

Resource	Technology	Century
Sun	Animal	<19th
Coal	Steam	~ 19th
Hydrocarbons	Combustion	~ 20th
Atoms	Electro-magnetic	~2?th
Outer space	Antigravity	~??th

As featured here, the technology of combustion is in principal place today. Confronted by the mounting dissatisfaction with current technology of internal combustion engines, the automotive industry is involved nowadays in a frantic search for alternative fuels and alternative energy conversion systems, with hydrogen fuel and PEM (Proton Exchange Membrane) fuel cell system at the focus of attention. There are many imponderables facing the decision makers in this connection. The fact that hydrogen is a superb fuel is well established. The mass of air per unit mass of hydrogen is ~ 34 - over twice that of a hydrocarbon – and, on top of that, the flame temperature at initially normal pressure and temperature is ~ 2300 K, whereas for a hydrocarbon/air mixture it is ~ 2100 K, both well above threshold of ~ 1800 K - the melting temperature of steel. Paradoxically enough, by the same token, the use of hydrogen involves a distinct hazard of fire, let alone explosion – a hazard exacerbated by the fact that its flammability range in air is very broad (the lean limit at normal pressure and temperature being at a level of ~ 1.5% in mass concentration), while its flame is of blue, practically invisible, color.

The hydrogen fed fuel cell is at the threshold between the combustion and electro-magnetic technologies. It manifests progress in the right direction – a step toward future. The question how much effort should be spent today on such an alternative system, while the attractive prospects of equivalent progress in the technology of combustion in piston engines are disregarded, is open.

   The purpose of this book is to provide the background for implementation of such a progress. This is accomplished by providing an assessment of the technology of combustion in piston engines, followed by furnishing an engineering method of approach for evaluation of the effectiveness with which fuel is utilized in the engine cylinder, and demonstrating how it can be improved. These subjects are exposed, respectively, in two parts: PART 1, SYNTHESIS and PART 2, ANALYSIS.

   Part 1 is made out of **Chap. 1 Overview**, describing what the technology of combustion in piston engines is all about, **Chap. 2 Perspective,** providing an account of how did it got to its present state, and **Chap. 3 Prospective,** pointing out how it technology can be advanced and what gains can be thereby derived.

   Part 2 is based on the recognition that combustion is at the heart of a piston engine. Ushered thus is a profession of engine cardiology. Chapters in this part are entitled, therefore, in medical terms: **Chapter 4 Diagnosis**, introduces the principles of this profession, **Chap. 5. Procedure**, prescribes the manner in which it is applied to a given engine (akin to examination of a patient), and **Chap. 6 Prognosis** provides an insight into its prospects. A practitioner of this profession is illustrated below by the portrait of Cal Bear playing this role, drawn by my friend, Jean-Pierre Petit*.

Jean Pierre Petit
1994

---

* <www.jp-petit.com>

Acknowledgement

I wish to express thanks to my colleagues, Profs. Ali Bulent Cambel, Alexandre Chorin, Herbert Heitland, Eiichi Murase, George Leitmann, Andy Packard, Cornel Stan, Harold Schock, and Erich Thomsen, to my students and associates, Johans Sum, Yuan Shen, Drs. Jean-Pierre Hathout, Jasim Ahmed and Aleksandar Kojic, as well to my great friend Motchek Openchowski, for greatly appreciated encouragements and comments, and, above all, to the grand companion of my life, Min, our greatly cherished daughter, Terry, and our magnificent grandchildren, Jessica and Zachary, without whose love, care and attention all that would not have been possible.

Contents

Part 1

PART 1

Synthesis

1 Overview

1.1 Purpose

The sole purpose of combustion in a reciprocating piston engine is to shift the expansion process away from the compression process in order to generate a working cycle. The only reason for the use of fuel is to generate pressure in order to accomplish this task. This is accomplished by an exothermic (one associated with acquisition of internal energy) chemical reaction, as a consequence of which a transformation (metamorphosis) takes place in engine cylinder between the reactants at initial state, i, attained at the end of the process of compression, and the products at final state, f, established at the start of the process of expansion, as depicted on the pressure-volume (indicator) diagram in Fig. 1.1.

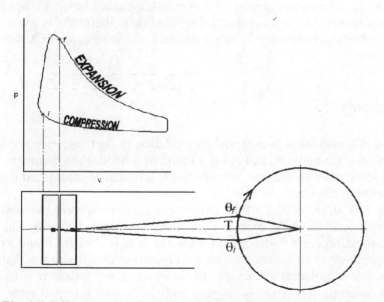

Fig. 1.1. Effective role of combustion in a reciprocating piston engine. *The dynamic stage from i to f, and its lifetime* $T = \Theta_f - \Theta_i$.

The directly evident outcome of this process, manifested by the measured pressure data, is referred to as the dynamic stage of combustion. Its lifetime, $T = \Theta_f - \Theta_i$, depicted on the crankshaft circle, is about 1/10 of a rotation. At a maximum speed of 6000 rpm, it is 1 millisecond in duration. At a cruising speed of 2000 rpm it is 3 milliseconds.

So, the fuel, with all the technological, industrial, economic, political and social aspects involved in its supply, accomplishes its effective action in a couple of milliseconds, faster than a blink of the eye – an essential moment of truth. It is also in the course of this short life of the dynamic stage that all the pollutants are formed. What happens there and then is thus of crucial importance to the operation of a piston engine. Described here are some of the fundamental features of this event.

1.2 Combustion

Combustion – the exothermic process associated with oxidation of a hydrocarbon fuel - is the oldest technology of mankind. Over the years this lead to a paradox. On one hand, combustion is approached with great reverence, like religion that has to be believed in without understanding. On the other hand, however, it is taken for granted as a consequence of intimate familiarity. The consequences of this paradox cannot be appreciated without some knowledge of this subject. Provided here, therefore, is a concise résumé of what combustion is about and how can it be best utilized in a piston engine.

1.2.1 Reactants

The process of combustion is executed by oxidation of fuel, accomplished by its chemical reaction with air. Fuel is a liquid or gaseous substance consisting of hydrocarbon molecules. A molecule is a bunch of atoms tied together by orbiting electrons.

There are four atoms of major relevance to combustion: carbon, denoted by C, and hydrogen, H, in the fuel; oxygen, O, and nitrogen, N, in air. The chemical expression for a hydrocarbon molecule is C_nH_m, where n and m denote, respectively, the number of atoms of carbon and hydrogen that make it up. For a molecule of octane, for example, n = 8, while m = 18. Air is predominantly a mixture of oxygen molecules with nitrogen molecules, expressed, respectively by O_2 and N_2, each consisting, evidently, of two atoms.

1.2.2 Pyrolysis and Dissociation

The chemical reaction of combustion is carried out in two main stages (1) molecular fission and (2) molecular transformation (metamorphosis). First, when heated up, the hydrocarbon molecules become agitated (activated). In an engine cylinder, this is executed either by an electric spark, or by compression. Powerplants utilizing the former are referred to as spark ignition engines, while those using the latter are diesel engines. As a consequence of the agitation, heavy hydrocarbon molecules lose hold of a hydrogen atom and break into lighter fractions referred to as radicals – an action called pyrolysis. As a consequence, the molecules get transformed into hydrocarbon radicals immersed in a "soup" of hydrogen atoms. Relatively light radicals can acquire thereby enough kinetic energy to become *active* in assisting further propagation of the process. Concomitantly, as a consequence of thermal heating, other molecules, such as oxygen, get also split into fractions – an action called dissociation. The processes of dissociation and pyrolysis occur at a temperature level < 1000 K[1].

1.2.3 Autocatalysis

Thereupon, the hydrogen atom, H – by far the lightest of all the atoms, and hence, the most agitated, the most mobile, the most ubiquitous – ushers in a chain reaction mechanism, playing thus a role of the most effective chain carrier among all other active radicals participating in this mechanism. By hitting (colliding with) an oxygen molecule, O_2, it gets an oxygen atom, O, split out, while combining itself with the other oxygen atom to form OH - a radical known as a hydroxyl. This event is described algebraically as $\underline{H} + O_2 = \underline{O} + \underline{OH}$, where symbols of active radicals are underlined. The oxygen atom, O, as well as the hydroxyl radical, OH, although, respectively, sixteen and seventeen times heavier than the hydrogen atom, play, nonetheless, roles of active radicals by getting involved in propagating the reaction further.

Thus, it is by such a chain reaction mechanism that the molecular transformation is accomplished. Each of the molecular collisions, of which it is made out, is an elementary step. The immediate consequences of initial agitation, such as pyrolysis and dissociation, are the initiation steps. The collisions of one chain carrier generating two chain carriers, as that of a hydrogen atom with an oxygen molecule illustrated above, is a chain branching step. Its threshold temperature is at a level of 1000 K.

---

[1] Kelvin, an absolute centigrade scale

Ushered in then are propagation steps, where a radical hits a molecule, combines with it, and causes another radical to split off. Eventually, the chain reaction mechanism of hydrocarbon oxidation ends with termination steps, associated with generation of oxygen-saturated molecules: carbon dioxide and hydrogen oxide (water).

The concept of chain reaction mechanism was formulated by Nikolai Nikolaievich Semenov, Nobel Prize Laureate and Director of the Institute of Chemical Physics in Moscow. It was conceived in the study of combustion upon the recognition that it is, in effect, auto catalytic in nature, providing thus the fundamental basis for the catalytic industry. He liked to refer to chain branching as an avalanche. In his eighties, he instituted a program of R&D, in association with automotive industry, to implement the chain reaction mechanism in an internal combustion engine. Its outcome was the so-called LAG[2] engine, featuring turbulent jet ignition, which attracted a good deal of attention among the automotive community in early 1980's.

1.2.4 Exothermicity

The oxygenated carbon and hydrogen molecules coming out of all these collisions at high velocities behave like a stack of billiard balls hit by a queue ball. Moreover they are rotating and vibrating. The molecules set into these mechanical motions acquire kinetic energies of three kinds: translational, rotational and vibrational. The sum of these energies is expressed as internal energy. The molecules, endowed with this energy, exhibit their agitated (activated) state by high temperature. A process of generating energized products at high temperature is referred to as exothermic. The site where all this takes place is referred to as an exothermic center.

1.2.5 Nitric Oxide

It is by impact of atoms that the molecules of oxygen and nitrogen get dissociated. At a high temperature (around 2000 K), this process forms nitric oxide, NO, one of the major pollutants. The thermal mechanism of generating NO is straightforward: an oxygen atom, O, hits a nitrogen molecule, N_2, yielding NO and O[3], while a nitrogen atom, N, hits an oxygen molecule, O_2, producing NO and O[4]. This mechanism was derived by one of the

---

[2] Lavina Activatsia Gorenia, i.e. Avalanche Activated Combustion
[3] $\underline{O} + N_2 = NO + \underline{N}$
[4] $\underline{N} + O_2 = NO + \underline{O}$

most prominent theoretical physicist in Russia, Yakov Borisovich Zeldovich, when he was a teenager.

1.2.6 Chemical Dynamics

Actually, an atom, unlike a billiard ball, is akin in its structure to a solar system of a star (proton) circumscribed by orbiting planets (electrons). A molecule is akin to a cluster of stars. Molecular collisions are, therefore, events of cataclysmic proportions. Their deterministic study, with a sophisticated theory of orbital dynamics supported by molecular beam experiments, is referred to as chemical dynamics – an eminent branch of science distinguished by a number of Nobel Prize laureates.

1.2.7 Chemical Kinetics

In the course of a single chemical reaction, there are many millions of such cataclysmic events taking place. Under such circumstances, one has to deal with millions of colliding worlds! Their eminently appropriate statistical interpretation is referred to as chemical kinetics. Thus, the science of chemical dynamics provides the fundamental background for the subject of chemical kinetics that treats all the atomic and molecular collisions on the basis of statistically established parameters. Collisions between two molecules are called bi-molecular. Collisions between three molecules are termolecular. The rate at which a chain reaction proceeds depends on the frequency of collisions for an elementary step to occur. The frequency is expressed statistically by probability. The probability of the occurrence of bimolecular collisions is, naturally, much higher than of termolecular collisions. To attain a high temperature, however, termolecular collisions are indispensable, because it is only the third partner that carries the gain in exothermic (elevating the temperature) energy generated by the chemical reaction. Their generation is tantamount to the termination steps. It turns out that the principal step for oxidation of carbon monoxide is a bimolecular collision: $CO + \underline{OH} = CO_2 + \underline{H}$, which is exceedingly slow. So slow, in fact, that it lags significantly behind the essential events of the exothermic reaction, depriving it of the most effective chain carrier, the hydrogen atom, $\underline{H}$, that could have a catalytic effect upon its evolution.

1.3 Flames

The course of nature is so universal that its outcome is amazingly similar under a wide variety of seemingly dissimilar circumstances. In particular, the resemblance of the consequences of chain reaction mechanism to the political events of our times is so uncannily fascinating, that it is utilized here to present the rest of the story. The agitated molecules behave as discontented agitators in a malcontent society. They spread out their agitated state at molecular speeds (about 20 % lower than the local speed of sound) - a process known as diffusion. This action has all the earmarks of shooting in the dark: it is carried out at random. The demonstration that random motion is the essential mechanism of molecular diffusion was cited as one of the accomplishments in the certificate of the Nobel Prize for Einstein. It is by this type of communication links that agitators form gangs, where the intensity of agitation can be made so cohesive and effective as to terrorize the rest of the world. In a combustion system left to itself, such as well organized gangs are flames – a set of exothermic centers tied together by molecular diffusion. Flames establish themselves in the form of thin sheets, across which the reactants are transformed (converted) rapidly into products. Their fronts are closed (continuous), delineating contours of the highest temperatures (degrees of agitation) achievable in the combustion field.

1.3.1 Pollutant Formation

A high intensity of agitation is sustained by a tight concentration of exothermic centers. As a consequence of the strong cohesion established by diffusion, once they are formed, it is practically impossible to break them apart. Thus, on one hand, flames are invariably formed when the process of combustion is left to itself to occur naturally, whereas, on the other hand, they maximize automatically the production of all the pollutants. This is so for the following reasons.

The exceedingly high temperature they achieve promotes the formation of nitric oxide. The flame fronts, where the action is, are so thin, that they cannot accomplish satisfactorily their mission of converting the reactants into products. It is for that reason that molecules of carbon monoxide, CO, are left out. Their oxidation into carbon dioxide, CO_2, is so slow that in a well-ordered flame known as laminar, it takes place well behind the front. As a consequence of this delay, the reaction is deprived of the most effective and, hence, the most valuable chain carrier, the hydrogen atom. To make matters worse, the cohesive flame fronts, established behind their diffusion layers, cannot penetrate all the nooks and crannies (cavities) of the cylinder-piston enclosure leaving a good chunk of hydrocarbon fuel completely unprocessed, the reason for an objectionable concentration of unburned fuel in exhaust gases.

Thus, flames are sources of all evil insofar as pollutant formation is concerned. Anything done to inhibit their establishment should go a long way in recovering from the drawbacks of pollutant formation and ineffective fuel utilization, as implied below and elucidated in Part 2.

1.3.2 Pollutant Abatement

It is the intimate familiarity with this essential nature of combustion that led to the concept of external treatment, associated with the technology of chemical processing plants in the exhaust pipe, like catalytic converters. Under influence of a moderately elevated temperature, a catalytic surface in these reactors, be it platinum or palladium, is a donor of oxygen atoms. As one of the chain carriers, this atom provides service in augmenting the rate of oxidation, accomplishing thus the mission of a catalytic agent. But the hydrogen atom does this significantly better and faster in the course of combustion in the cylinder. It acts then, *de facto,* as a catalytic agent, for, as pointed out in the previous section, combustion reaction is intrinsically auto catalytic. Since hydrogen atoms, generated in the course of interior treatment, are much more effective than oxygen atoms provided by surface catalysis of external treatment, the former is fundamentally superior to the latter.

1.4 Knock

The phenomenon of knock played a crucial role in the evolution of the technology of combustion in piston engines. In the past, it was, therefore, of central significance to both, the auto- and the oil-industries. Today, its solution is well known to any motorist in terms of the octane rating of gasoline displayed at all service stations. A similar criterion, in terms of the cetane number, applies to diesel oil. The fact that this is associated with a significant increase in the cost of fuel is accepted as a *fait accompli.*

Upon significant research studies, carried out to explore the physics of detonation phenomena and explosions, it has been established that knock is, in effect, a blast wave[5] generated by a concentrated deposition of energy in a gas at a high rate. When its shock front interacts with the piston face, it acquires the action of a chisel that hits it, inflicting structural damage, rather than exerting pressure to push. The principal way to advance the technology of combustion in piston engines, advocated here, is associated

---

[5] Transient flow field emanating from source at, or around, the center and bounded by a shock front at its periphery

intimately with dilution attained by mixing fuel with extra air, as well as with recirculated residual or exhaust gas. Thus, the localized rate of energy deposition (power density) is concomitantly reduced to such an extent that the onset of knock is annihilated. The octane rating of gasoline can be thereby eliminated – an achievement that can yield an appreciable reduction in the production cost of fuel.

1.5 Prospects

In real life, correcting a misfortune is very difficult, if not entirely impossible. The same applies to combustion. As pointed out above, once a flame is established, it is practically impossible to break it, as long as the reactants are available. In a piston engine, however, one has a multiple sequence of lives. For a car traveling at an average speed, thousands of reincarnations take place per minute. With assistance of modern computer and control technology, any mishap can be, therefore, prevented from occurring again within the time interval of tens of milliseconds available between these reincarnations.

As it should be apparent by now, the main sources of evil are flames, where the exothermic centers are ganged together to form a highly agitated front. This concentrated action is depicted in Fig. 1.2, where the Cal Bears, representing exothermic centers, are forced to cultivate a field by getting bunched together in an overcrowded row at the front.

Fig. 1.2. A propagating flame front illustrated by a row of a gang of morose Cal Bears forced to cultivate a barren field in a tight line; drawn by Jean Pierre Petit[*]

---

[*] <www.jp-petit.com>

Fig. 1.3. Distributed reaction centers illustrated by happy Cal Bears, each content to work on his own, while the field is cultivated throughout its full extent, demonstrating the advantage of action in parallel rather than in series. Drawing of Jean Pierre Petit, with portraits of his favorite characters, Monsieur Anselme Lanturlu and his girlfriend, Sofie, with their two wisecracking birds providing voluble comments on what's going on

An alternative way to accomplish the task is illustrated by Fig. 1.3, where the Cal Bears are distributed, each having fun in cultivating his individual flowerpot. The cultivation, performed painstakingly in series by a propagating front of malcontent bears, is accomplished in parallel much better and faster when they are happy for being treated as well as they ought to be. Therefore, instead of permitting flames to be established beyond any control, one should not let exothermic centers to agglomerate into delinquent gang-lines.

In an engine cylinder adequate distribution of exothermic centers to prevent the formation of flames can be attained by dilution. This can be accomplished either by pre-mixing fuel and air using RGR[6] or EGR[7], or injecting turbulent jets of super-rich air/fuel mixture into a piston-compressed air combined with a certain amount of RGR or EGR. The concept of premixing the combustible charge with a relatively large amount of residual gases gained today quite a lot of popularity under the name of HCCI[8]. The adjective "Homogeneous" implies in reality the distribution of exothermic centers portrayed by Fig. 1.3, whereas "Compression Ignition" is handicapped by uncontrollability. However, dilution of the charge offers remarkable means to inhibit the occurrence of knock as a consequence of the diminished concentration and amplitude of exothermic centers, reflected by a decrease in local exothermic power density.

---

[6] Residual Gas Recirculation
[7] Exhaust Gas Recirculation
[8] Homogeneous Charge Compression Ignition

It is quite reasonable to expect that the execution of the exothermic process could be monitored and modulated for this purpose by a micro-electronic control system of a CCE[9]. Conceptually, the electronic apparatus consisting of micro-sensors for monitoring pressure and MEMS[10]-type devices for operating its micro-valves could be accommodated within a 'smart cylinder head' that, eventually, may be miniaturized to a "smart cylinder gasket". The control system will be governed by a microprocessor. As pointed out above, its task will have to be accomplished for each combustion event within a time interval of an order of a millisecond between two consecutive cycles – well within the capability of modern electronics.

To sum up, the secret of success in controlling combustion is based on the political adage of Julius Caesar: DIVIDE ET IMPERA.

---

[9] Controlled Combustion Engine
[10] Micro-Electro-Mechanical-Systems

2 Perspective

2.1 Background

The literature on processes taking place in the cylinder of an internal combustion engine is very rich, indeed, as should be expected of a topic bearing so incisively upon the automobile and oil industries that occupy today a most prominent economic sector in the world. This is manifested, of course, by the textbooks on piston engines[1], as well as by classical monographs on combustion[2].

The impressive status quo in the technology of piston engines at which it is today has been attained as a consequence of many contributions from the academic and industrial research institutions; too many, in fact, to be acknowledged here. As a sample of academe, consider just the legacy of Technische Hochschule Aachen established by F.A.F. Schmidt and Franz Pischinger, Massachussetts Institute of Technology under the guidance of Fayette Taylor, John Heywood and Jim Keck, the University of Wisconsin under the leadership of Philip Myers and Otto Uyehara. Representing prominently R & D organizations are Ricardo Engineers in Shoreham-by-Sea, England, founded by Sir Harry Ricardo, AVL[3] in Graz, Austria, founded by Hans List, FEV[4] in Aachen, Germany, founded by Franz Pischinger, as well as JARI[5], founded in 1969 as a "non-profit organization for promoting a healthy progress of motorization in society at large".

The traditional approach to the theory of combustion in piston engines is concerned with the direct problem of evaluating the outcome from the basic principles of thermodynamics, heat transfer chemical kinetics and fluid mechanics. In recognition that the sole purpose of combustion in engines is to generate pressure, presented here in Part 2 is a method of approach to

---

[1] Ricardo 1922–1923; Saas 1929; List 1939–1946; Pischinger and Cordier 1948; Schmidt 1951; 1965; Benson 1962; Taylor 1982–1985; Obert 1973; Heywood 1988; Horlock and Winterbone 1986; Pischinger et al. 1989–2002; Blair 1996, 1999

[2] Jost 1946; Lewis and von Elbe 1987; Williams 1985; Zeldovich et al. 1990

[3] Anstalt für Verbrennungskraftnaschinen Prof. Dr. Hans List

[4] Forschungsanstalt für Energie und Verbrennung

[5] Japan Automobile Research Institute

the inverse problem of evaluating the action taking place in the cylinder, from its outcome recorded by the pressure transducer measurement. Particular attention is restricted then to the essential part of the process of combustion when it creates pressure pushing the expansion process away from the compression process, referred to as the *dynamic stage*.

2.2 Milestones

To consider the future of a technology, an appreciation of its evolution is a condition *sine qua non*. Provided here, therefore, is a historical review of the milestones of progress in the execution of the exothermic process of combustion in piston engines that ushered in the status quo. Historically, the major obstacle in the development of internal combustion engines was the phenomenon of knock. Consequently, major R&D effort to prevent its occurrence was at the forefront of initial progress.

Fig. 3 – 1920 car in which compression ratio was boosted from 4 to 1 up to 7 to 1

Fig. 2.1. First milestone in the technology of combustion in piston engines (Boyd 1950)

2.2.1 Quality of Fuel

A concept introduced by Charles F. Kettering, who, using a variable compression engine developed by C.F Taylor, launched the technology, of fuel additives that plays still a prominent role in automotive industry.

Its discovery was made in Dayton, Ohio, the venue of the Wright Pat-terson Air Force Base, and a colorful history of this event was recounted by Kettering's associate, T.A. Boyd (1950). The first milestone in the technology of combustion in piston engines was thus laid down, as pointed out triumphantly in the caption of his Fig. 3 reproduced here as Fig. 2.1. The national Committee for Fuel Research, formed thereupon, introduced the fuel rating standards, the octane number and, later, the cetane number – the criteria in force today throughout the world.

2.2.2 Geometry of Combustion Chamber

A concept pioneered by Sir Harry Ricardo (1922–1923), the founder of a world-renowned engine research laboratory in England. Upon the belief that knock is triggered by pre-ignition of the "end gas", the cylinder head introduced by him was shaped so as to provide most of the surface area to the combustible mixture away from the ignition spark for its cooling to quench its explosive tendency.

2.2.3 Fuel Injection

Fuel injection is a characteristic feature of diesel engines. In early times, this was accomplished by the use of compressed air. Consequently, all die-sel engines had to be outfitted then with auxiliary compressors. The major breakthrough was achieved towards the end of nineteen twenties by the in-troduction of 'solid injectors,' which promoted the technology of precision machining, as brought out emphatically by the title of his book, "*Kompres-sorlose Dieselmaschinen*" by Friedrich Saas (1929). Nowadays, all the diesel engines feature "solid fuel injection", featuring either separate (di-vided) combustion chambers, with or without cavities in the cylinder into which the fuel spray is directed, or direct injection. – the technology of modern diesel engines.

2.2.4 Quality of Combustion

A concept introduced by Rassweiler and Withrow (1938). Upon revealing the mechanism of knock in its various forms by means of high speed cinematography, (Withrow and Rassweiler 1936), they used the same pho-tographic apparatus, concomitantly with measurement of the pressure pro-file, to observe the evolution of the flame motion as it traverses the charge in the engine cylinder, and deduced on this basis an analytical technique for cycle analysis to assess the influence of fuel quality upon engine performance. The technique they thus established is still in common use today as the "heat release analysis".

2.2.5 Catalytic Fuel Refinement

A concept introduced in the mid nineteen thirties for producing gasoline of high-octane number by catalytic refinement, known as the Houdry process. The product of it, in demand at that time for aircraft engines, formed one of the major items shipped to Great Britain within the ramification of the Lend-Lease program established with the USA at the start of the Second World War. This technology is the forerunner of the catalytic hydrogen reformation employed today.

2.2.6 Jet Ignition

A concept introduced by Nikolai Nikolaievich Semenov (1958–1959), the Nobel Prize Laureate for formulating the chain reaction theory. Upon over half a century of an extensive R&D program he directed with Lev Ivanovich Gussak (1976, 1983), the first jet ignition engine was produced under the name of LAG (for Lavina Activatsia Gorenia, i.e. Avalanche Activated Combustion).

2.2.7 Flameless Combustion

A concept first conceived by Shigeru Onishi, who turned his annoyance with "diesling" of his motorcycle engine into a profitable venture by, first, learning more about it in association with Satoshi Kato, who became thereupon his chief engineer, and, then, developing it into a manufacturing enterprise (Onishi et al. 1979). At the same time Noguchi carried out a thorough study of this mode of combustion at Toyota Motor Co. in collaboration with Nippon Soken (Noguchi et al.. 1979). Four years later, Najt and Foster (1983) demonstrated that the same type of combustion can be also achieved in a four-stroke engine, and referred to it as HCCI[6] – a name by which it is known today.

---

[6] Homogeneous Charge Compression Ignition

2.2.8 Gasoline Direct Injection

A concept, developed initially for diesels, introduced recently to gasoline engines upon an extensive R&D effort. A comprehensive account of its current state of art has been provided by Cornel Stan (2000, 2003), including the rendition of direct injection systems in their early stages developed for diesel engines. Its essential feature is the execution of the exothermic process of combustion in sprays, without any preliminary steps for initiating the process of combustion employed before under the name of "indirect injection", such as pre-chambers, indentations in piston crowns, cavities in the cylinder known as "energy cells", etc. The recently attained success in this field was instrumental in renewing the technology of open chamber stratified charge engines, abandoned over twenty years ago by the open chamber stratified charge engine projects of TEXACO, PROCO of Ford, and DISC of GM.

2.3 Implementation

Salient features of recent milestones are henceforth systematically reviewed in an order in which, it is expected, they will be implemented by the automotive industry.

2.3.1 Direct Injection

The concept of direct injection is well established in automotive technology. Its modern use in diesel engines was chiefly responsible for the advanced status they acquired among automotive powerplants. The adaptation of this concept to gasoline engines ushers in a way into the no-man's land between the two, offering an attractive prospect of fuel independence.

The technological progress in this respect was supported by an impressive amount of studies on spray dynamics, described, among others, by of Arthur Lefebvre (1989) and William A. Sirignano (1999).

2.3.2 Distributed Combustion

Insofar as the formation of pollutants is concerned, flames are the source of all evil, as brought out in Chap. 1.

The reason for it is that, by their intrinsic nature, they are established at stoichiometric[7] contours where the highest temperature, achievable in a

---

[7] Appropriate chemical composition of the air/fuel mixture producing fully oxidized products

given fuel/air mixture, is attained, enhancing the formation of nitric oxide, while the rate of the exothermic reaction is so high that the residence time of the reacting substance is far too short for an adequate oxidation of the hydrocarbon fuel and carbon monoxide. It is in this way that flames create optimum conditions for the generation of all the legally identified pollutants. Inhibiting the formation of flames by dilution associated with RGR, exemplified by HCCI engines, is, therefore, a pioneering achievement. Besides bridging the gap between spark and compression ignition engines, it is, in fact, the first attempt at an intrusive interference, rather than fuel quality and additives, into the execution of the exothermic process of combustion in an engine cylinder.

The experimental setup employed by Onishi et al. (1979). is depicted in Fig. 2.2.

The results of their studies are displayed by Figs. 2.3 and 2,4. The first, a sequences of Schlieren[8] records of the exothermic process of combustion, demonstrates the achievement of flameless combustion by means of adequate RGR – a method they called, for lack of a better name, ATAC[9], without much meaning attached to it. The second provides evidence of the miraculous effect of distributed combustion promoted by RGR in 'miraculously' annihilating the cycle-to-cycle variations.

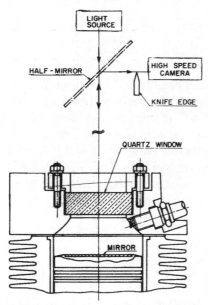

Fig.2.2. Experimental engine set up of Onishi et al. (1979)

---

[8] Optical technique for visualization of sharp density gradients
[9] Active Thermo-Atmospheric Combustion

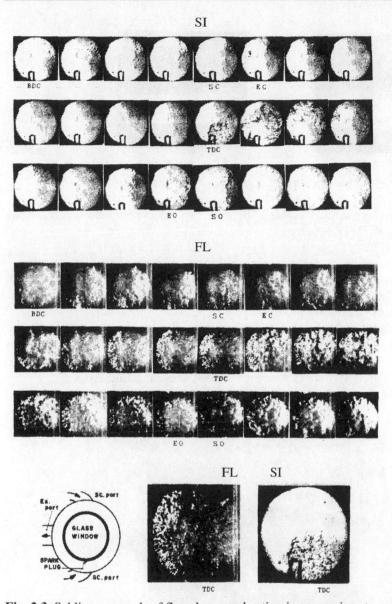

Fig. 2.3. Schlieren records of flameless combustion in comparison to conventional spark ignited flame according to Onishi et al. (1979)
SI *spark ignited combustion;* FL *flameless combustion referred to as ATAC*

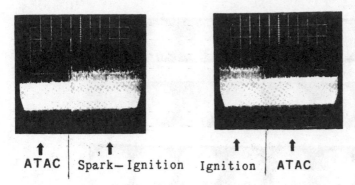

$\uparrow$ | $\uparrow$ $\uparrow$ | $\uparrow$
ATAC | Spark—Ignition   Ignition | **ATAC**

Fig. 2.4. Peak pressures in switching between two modes of combustion according to Onishi et al. (1979)

A commercial product of these studies is presented by Fig. 2.5: a photograph of the: an ATAC engine driven generator providing 1 kW for individual power generation in Japan.

Appreciating the advantages achieved by dilution of the air/fuel charge with RGR, demonstrated by Onishi, Masaaki Noguchi [23] launched an impressive program of research using a two-stroke opposed-pistons engine equipped with ample spectroscopic sensors, shown in Fig. 2.6

Fig. 2.5. Portable NICE-10GC generator (Onishi et al. 1979)

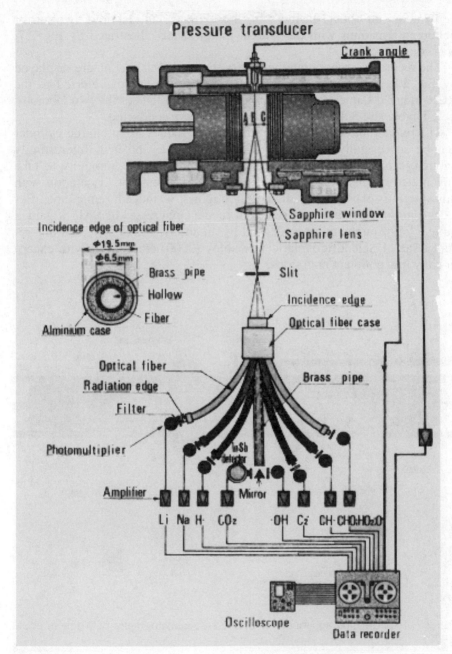

Fig. 2.6. Experimental apparatus for spectroscopic and pressure measurements of Noguchi et al. (1979)

To attain an optical insight into the cylinder, an elaborate two-mirror Schlieren apparatus with transparent piston heads, illustrated by Fig. 2.7, was constructed.

The records corroborating the observations of Onishi et al. are displayed by Fig. 2.8. Here again, for lack of a better name, they referred to the RGR-induced flameless combustion as TS, the initials of the two laboratories, Toyota and Soken where these studies were conducted.

The remarkable fact that the formation of flames in an engine cylinder can be successfully inhibited was thus firmly established. Interestingly enough, it took over twenty years for the automotive community to take any notice of this pioneering achievement. The current fascination with this system is expressed by the HCCI[6] engines, is, indeed, remarkable. Besides a number of papers presented at recent Congresses of SAE, it gained quite a lot of publicity, as evidenced by an article, which appeared in a recent issue of Scientific American (Ashley (2000) crediting it with exceptionally low pollutant emissions.

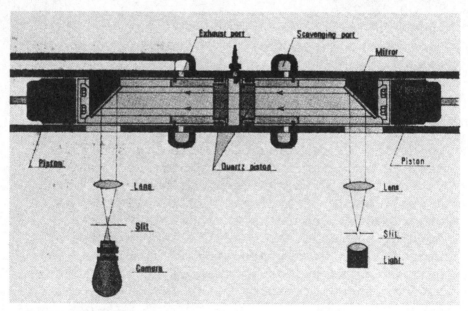

Fig. 2.7. Experimental apparatus for Schlieren cinematography of Noguchi et al. (1979)

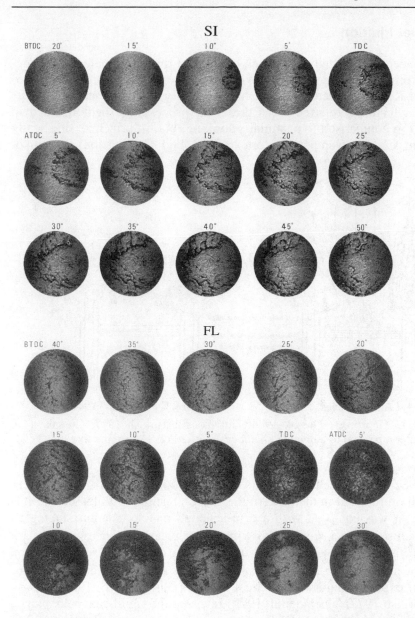

Fig. 2.8. Schlieren records of spark-ignited combustion in comparison to flameless according to Noguchi et al. (1979)
SI *spark ignited combustion;* FL *flameless combustion referred to as TS*

2.3.3 Jet Ignition

The concept of jet ignition is much older than that of flameless, homogeneous combustion due to RGR, having been an object of a significant program of R&D, conducted for over half a century, on flame jet ignition. Its essential features, referred to as the LAG process, are demonstrated in Fig. 2.9 in comparison to a similarly looking, but, in principle, drastically different CVCC (Compound Vortex Controlled Combustion) system of Honda.

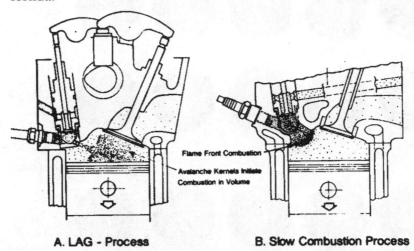

Flame Front Combustion

Avalanche Kernels Initiate
Combustion in Volume

A. LAG - Process   **B. Slow Combustion Process**

Fig. 2.9. LAG process (Gussak 1976; Gussak and Turkish 1977; Gussak et al. 1979) in comparison to a divided chamber stratified charge engine like the CVCC of Honda

In 1981, upon some twenty years of tests in military trucks, it was ostentatiously introduced in the powertrain of the prestigious Volga car. In this version, it was equipped with a cam-actuated injector for introducing an extra-rich ($\lambda \approx 0.5$) air/fuel mixture into a pre-chamber, where, upon spark ignition, a turbulent jet is issued, providing a distributed set of ignition sites for an ultra-lean ($\lambda \approx 2$) mixture in the cylinder. An extensive program of experimental and theoretical studies (Gussak 1976; 1983; Gussak and Turkish 1977; Gussak et al. 1979) revealed the significant role, played in this process by active radicals.

An interesting attempt to obviate the cumbersome mechanically actuated injection was made subsequently by Reinhard Latsch (1984) at Bosch, Suttgart. In this case that was achieved by relying on admission of only the piston compressed air/fuel mixture to the igniter cavity, eliminating thus the need for additional supply of its feedstock from an outside source. His flame jet igniter, introduced under the name of Swirl-Chamber Spark Plug, had this cavity miniaturized so that it could be accommodated within the body of a 14 mm spark plug.

The technology of jet igniters was advanced significantly by studies of plasma jet ignition conducted at the University of California[10] in cooperation with Dale, Smy and Clements at the University of Alberta[11], followed by a comprehensive program of research on flame jet ignition, carried out at the University of California[12], and continued in association with Kyushu University[13].

Plasma jet ignition (PJI) differs from flame jet ignition (FJI) for the following reasons. PJI operates upon a discharge of ~ 1 kV, associated with a relatively high current to generate plasma from any gas contained in a cavity of ~ 10 mm^3 in volume. Produced thus is a supersonic plasma jet creating a plume of extremely high temperature and relatively small mass, associated with an insignificant effect upon turbulent mixing. On the other hand, FJI operates upon a discharge of ~ 20 kV, associated with negligible current, to ignite an extra rich air/fuel mixture ($\lambda \approx 0.5$) contained in a cavity volume of $0.15 - 0.5$ cm^3.

It generates a subsonic turbulent flame jet of sufficiently large mass to induce mixing by entrainment of the combustible charge, forming a distributed turbulent combustion plume. Subsonic flame jet ignition is, of course, significantly slower than that obtainable by plasma jets. For this reason, for engines operating at higher rotational speeds, this technology of jet ignition should be developed further as a cross-breed between flame and plasma jet, as advocated in the next chapter. An example of a flame jet igniter is provided by Fig. 2.10.

To illustrate the performance of FJI generated combustion, in contrast to a flame, displayed here in Figs. 2.11-2.15 are five sets of cinematographic Schlieren records. The test vessel used for this purpose was a closed cylinder 3.5" (8.9 cm) in diameter and 2" (5.1 cm) deep, amounting to 315 cm^3 in volume. It was, in effect, the size of a CFR[14] engine cylinder at a compression ratio of 8 : 1, with piston is at 60 degrees crank angle from the top dead center (TDC). For each test it was filled with a carefully premixed propane-air mixture of 0.6 in equivalence ratio, initially at a pressure of 5 bars and a temperature of 65 °C. The mass of the combustion system was $M_s = 1.49$ gm.

---

[10] Oppenheim et al. 1978; Cetegen et al. 1980; Edwards et al. 1983; Edwards et al. 1985

[11] Dale and Oppenheim 1981; Smy et al. 1997

[12] Oppenheim et al. 1977, Oppenheim et al. 1989; Oppenheim et al. 1990; Maxson et al. 1991; Hensinger et al. 1992

[13] Murase et al. 1994; Murase et al. 1996

[14] Committee for Fuel Research

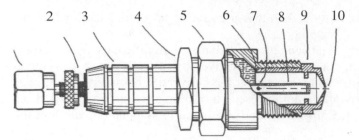

Fig. 2.10. Flame jet igniter (Oppenheim et al.1989; Oppenheim et al. 1990; Maxson et al. 1991; Hensinger et al. 1992)
1-fastener for insulator conduit delivering extra-rich air-fuel mixture; *2*-fastener for high-voltage trigger pulse; *3*-insulator body; *4*-fastener of insulator body; *5*-fastener of jet igniter; *6*-base of the insulator body; *7*-perforation for admission of charge into the generator cavity; *8*-delivery tube for the generator charge with closed end; *9*-ignition electrode; *10*-jet forming orifice

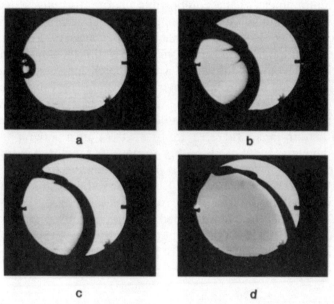

Fig. 2.11. Sequence of Schlieren records of combustion in a spark-ignited FTC (Flame traversing the charge) (Hensinger et al. 1992)

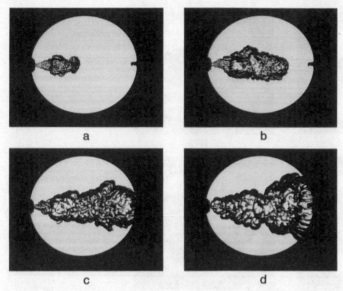

Fig. 2.12. Sequence of Schlieren records of combustion in a jet ignited SFMC (Single-jet Fireball Mode of Combustion) (Hensinger et al. 1992)

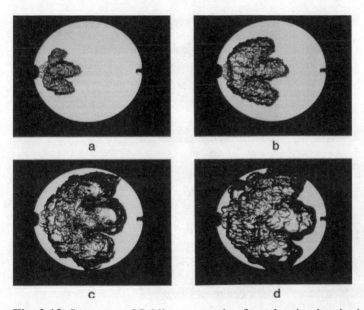

Fig. 2.13. Sequence of Schlieren records of combustion in a jet-ignited TFMC (Triple-jet Fireball Mode of Combustion) (Hensinger et al. 1992)

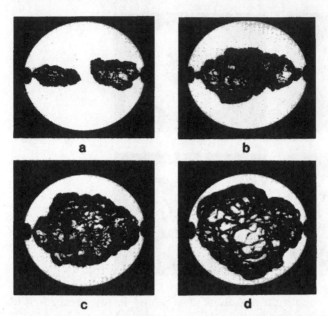

Fig. 2.14. Sequence of Schlieren records of combustion in jet-ignited OSFMC (Opposed Single-jet Fireball Mode of Combustion) (Hensinger et al. 1992)

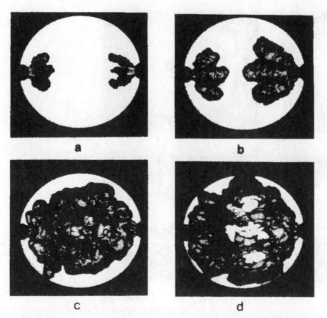

Fig. 2.15. Sequence of Schlieren records of combustion in jet-ignited OTFMC (Opposed Triple-jet Fireball Mode of Combustion) (Hensinger et al. 1992)

High-speed Schlieren cinematographs were taken through optical plates closing the cylinder at both ends. By the use of a variety of ignition systems, the experiments covered a wide range of operating conditions, from an apparently laminar flame to a fully developed turbulent combustion.

The five different modes of combustion were achieved by the following techniques in which the exothermic process of combustion was executed: (1) Spark ignited flame traversing the charge, F, Fig. 2.11. (2) Single stream flame jet initiated combustion, S, Fig. 2.12. (3) Triple stream flame jet combustion, T, Fig. 2.13. (4) Single stream opposed flame jets initiated combustion, OS, Fig. 2.14. (5) Triple stream opposed flame jet combustion, OT, Fig. 2.15.

In mode F, the flame front was clearly delineated, acting as the boundary between the reactants and products. The jet plume in mode S appeared as a round turbulent jet plume, while that in mode T was disk-shaped, in better match with the geometry of the enclosure. Both of these phenomena were enhanced by the use of opposed flame jets. As evident from the photographs, in mode F the combustion products were in contact with the walls right from the outset. In mode S, the combustion zone reached the walls after a distinct time delay, while in mode T this delay was distinctly longer. The contact of combustion products with the walls was made progressively shorter by the use of opposed single and triple jet configurations to create fireballs.

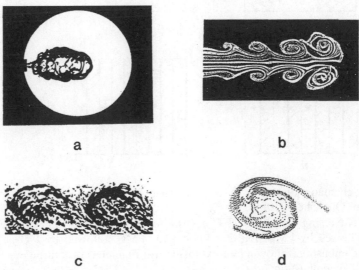

a

b

c

d

Fig. 2.16. Anatomy of a turbulent jet plume

To reveal the mechanism of a turbulent jet plume, its structure is portrayed in Fig. 2.16. Shown in **a** is a schlieren photograph of the plume soon upon its formation, like that of Fig. 2.12. Reproduced in **b** is a classical photograph of a jet plume in liquid demonstrating its large vortex structure. Provided in **c** is a close-up of a schlieren record of a turbulent plume illustrating its vortex nature. Depicted in **d** is an output of a CFD analysis of the flow field generated by a turbulent jet. Brought up thus is the essential role played by shear-generated vortices in a turbulent jet plume.

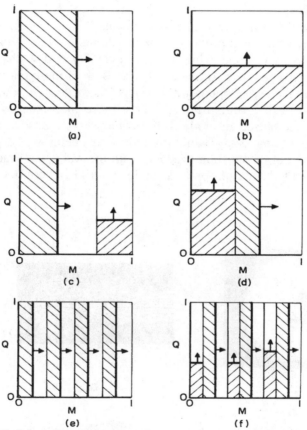

Fig. 2.17. Schematic diagrams of two alternative propagation mechanisms of combustion (Oppenheim 1984).
(a) Flame Traversing the Charge (FTC); **(b)** Fireball mode of Combustion (FMC); **(c)** FTC followed by FMC: end-gas knock; **(d)** FMC followed by FTC: "afterburn", i.e. combustion following an explosion; **(e)** Flamelets of a turbulent flame; **(f)** Distributed combustion
M expresses mass fraction of the cylinder charge; Q denotes energy fraction deposited by the exothermic process in the cylinder charge

Figure 2.17 provides a schematic description of two essentially different propagation mechanisms of combustion: FTC (Flame Traversing the Charge) and FMC (Fireball Mode of Combustion) in various configurations.

Their essential features are exposed by (**a**) and (**b**). Illustrated by (**c**) is the case of FTC followed by FMC - the essence of the end-gas knock, while (**d**) depicts the reverse: an FMC followed by FTC – a so-called 'afterburn,' that occurs upon explosion of a chemical charge (like TNT) in air. Displayed in (**e**) are multiples of FTC's and in (**f**) FMC's followed by FTC's, the first demonstrating the essence of what is known as "flamelets" of a turbulent flame, and the second bringing out the intrinsic features of distributed combustion.

3 Prospective

3.1 Background

We are living today in exciting times of spectacular advances in high-tech - a progress highlighted by the advent of micro-electronic control systems of nanosecond resolution. So far, the execution of the exothermic process of combustion (known, unfortunately, under a misleading name of "heat release") has been considered (appropriately for this misnomer) to be beyond any control. As a consequence of this misconception, the automotive industry acquired the practice of a "hands-off" technology of combustion. Everything possible is done to refine the fuel, provide it with more-or-less useful additives, and igniting it, either by spark or by compression, but, once the fuel is ignited, the rest is considered to be in the hands of God. The question one may pose is what can be done to foster progress and what can be thereby achieved (Oppenheim 2002).

3.2 Concepts

In view of the above, it seems obvious that progress in this field can be achieved by the development of "hands-on" technology – an advancement associated with the introduction of Microtechnology, in accordance with the best that today's high-tech can offer. One can expect its evolution to be based, of course, on the current state of the art, and the advances made step-by-step in pace with progress in the technology of electronic control. Salient features of developments to be made are presented here in the same order in which their status quo was reviewed in the previous chapter, namely: (1) direct jet injection (2) distributed combustion, (3) jet ignition, with a notable addition of (4) controlled combustion.

3.2.1. Direct Injection

The concept of direct injection, well established in diesel engines, has been recently extended to light fuel, gasoline engines with remarkable success described in Chap. 2.

The future of this technology is implied by combustors for gas turbine engines, which, right from the outset, were associated intrinsically with direct fuel injection. At the start, this involved secondary air mixing for reducing the high temperature attained at the stoichiometric primary zone where fuel was injected, to protect the turbine blades from melting. Produced thereby was a considerable amount of pollutants associated with significant emissions of unburned hydrocarbons and soot. The quality of combustion was greatly improved by the introduction of air-blast atomizers for injection - a technology significantly different from that of air-forced (or pressure) injection, employed in early diesel engines, using high-pressure air to push out liquid fuel into the cylinder and, thereupon, form a spray caused by the (Rayleigh) instability of the jet. In an air-blast atomizer, developed by Arthur Lefebvre (1989; 1983), microscopic fuel droplets are sheared off a liquid sheet of fuel within the body of the injector by air stream at a relatively low pressure. Its turbulent plume is capable of entraining a sufficient amount of air to create a stratified charge of a combustible mixture, where the droplets are evaporated and mixed before they reach the zone of exothermic reaction.

A conceptual design of a pulsed air-blast atomizer for diesel engines is depicted on Fig. 3.1 (Oppenheim et al. 1990). For a gaseous fuel – the sole kind admissible by fuel cells – liquid atomization is, of course, unnecessary; the turbulent jet plume of an air fuel mixture is then generated directly by a mixing valve.

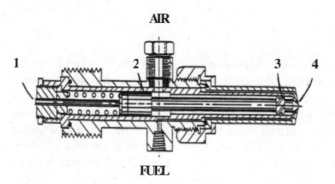

Fig. 3.1. Conceptual design of a pulsed air-blast atomizer (Oppenheim et al. 1990). *1*-actuator; *2*-pintle; *3*-swirl guide; *4*-atomizing nozzle

Whereas, in a gas turbine combustor, this system is operated at a steady rate, for piston engines it has to be implemented in an intermittent mode. Although principles of such an air-blast atomizer have been formulated some time ago (Dale and Oppenheim 1982; Oppenheim 1992, 1998; Oppenheim and Kuhl 1995), their engineering application is yet to be accomplished. It is of interest to note that the use of a turbulent air jet to enhance the burning rate is appreciated by the automotive industry, as exemplified by its recent adaptation to a prototype piston engine by SAAB (Olofsson et al. 2001).

3.2.2 Distributed Combustion

The concept of bridging the gap between spark ignition and diesel engines has been manifested recently by the rekindled interest in the HCCI[1], engines. Its major advantage is realized by impeding the formation of flames – the extremely narrow regions of the exothermic process of combustion where the maximum temperature is attained, while the residence time of the reacting substance is relatively short. As a consequence of the high temperature achieved in flames, the formation of NO is enhanced. Concomitantly, because of the short residence time of the reactive substance in them, little chance is left for carbon to be fully oxidized, enhancing the high concentration of CO and unburned fuel. So getting rid of flames is certainly a significant factor in the progress of combustion technology. The major disadvantage of HCCI is that the execution of the exothermic process of combustion is then essentially uncontrolled in both, time and space.

3.2.3 Jet ignition

A significant improvement of the HCCI concept can be made by getting rid of its major disadvantage: uncontrollability. The control has to be provided by a monitored supply of properly timed, distributed ignition sites. This can be achieved by micro-electronically controlled jet injection to create turbulent plumes, made out of large-scale vortex elements ("blobs"), where the exothermic process of combustion is initiated by jet ignition to provide a multiplicity of well distributed ignition kernels (vid. Figs 2.12–2.17). In familiar terms, the outcome is a properly modulated open-chamber stratified-charge engine. In the past, systems of this kind, like TEXACO, PROCO, and DISC, were handicapped by the use of concentrated, rather than distributed, ignition sources for intrinsically uncontrolled liquid sprays enhancing the formation of flames.

---

[1] Homogeneous Charge Compression Ignition

3.2.4 Controlled Combustion

The key to success in the implementation of these concepts is the development of a robust, feedback and feed-forward, micro-electronic control of combustion system, referred to as MECC[2], operating a PJI&I[3] system to create intermittently turbulent combustion plumes for execution of the exothermic process (Oppenheim 1992, 1998; 2002; Oppenheim and Kuhl 1995; Oppenheim et al. 1996). To provide an idea of its kind, illustrated here in Figs. 3.2 and 3.3 are its salient features. Since the essential purpose of this engine is to inhibit the formation of pollutants, it has to operate with an ultra-lean mixture. To compete in power output with four stroke engines, in prevalent use today, it should be a two-stroke, low friction engine, so that it could operate at high speed. The elementary components of a micro-electronic CCU[4] are depicted in Figs. 3.4–3.7. The first three provide schematic diagrams of, respectively, a fuel supply module for mixing air with fuel in prescribed proportions, a jet injector and jet igniter actuators. The last demonstrates the structure of the micro-electronic governor and its configuration with an ECU[5].

In a MECC engine, a controllable amount of RGR[6] will be provided and a controllable, in shape and time, plume of combustible mixture will be produced by multi-jet injection away from the walls, to protect it from their detrimental effects. Thereupon, the plume will be ignited by a number of flame jets, issuing inward from MEMS[7] valves situated on the periphery of the cylinder, in the vicinity of its head.

Part-load operation of such an engine will be controlled by adjusting the air/fuel mixture in the plume at WOT[8]. In contrast to the lack of controllability and flexibility of HCCI, MECC system will be able to govern the execution of the exothermic process of combustion in a more controllable and flexible manner than any piston engine available today. It is quite reasonable to expect that it should be possible to accommodate the entire MECC system in a "smart cylinder head" that, eventually, could be miniaturized into a smart cylinder gasket.

---

[2] Micro-Electronically Controlled Combustion
[3] Pulse Jet Injection and Ignition
[4] Combustion Control Unit
[5] Engine Control Unit
[6] Residual Gas Recirculation
[7] Micro-Electro-Mechanical System
[8] Wide Open Throttle

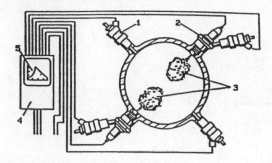

Fig. 3.2. Smart cylinder head (SCH) of MECC operating a PJI&I system (Oppenheim and Kuhl 1995)
1-injector; *2*-flame jet generator; *3*-turbulent jet plumes; *4*-electronic control modulator; *5*-pressure transducer and shaft decoder records portrayed by an indicator diagram

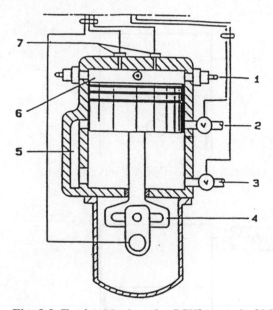

Fig. 3.3. Engine block under SCH[9] control of MECC operating a PJI & I system (Oppenheim and Kuhl 1995)
1-flame jet generator; *2*-tunable exhaust duct; *3*-tunable inlet duct; *4*-scotch yoke transmission; *5*-charge transfer duct; *6*-disk combustion chamber; *7*-pressure and temperature sensors

---

[9] Smart Cylinder Head

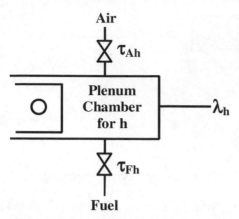

Fig. 3.4. Schematic diagram of a fuel supply module (FSM) (Oppenheim et al. 1996)
h: **Ij** for injector, **Ig** for igniter

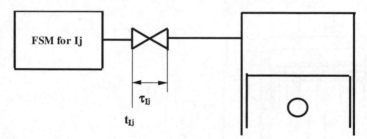

Fig. 3.5. Schematic diagram of a jet injector module (Oppenheim et al. 1996)

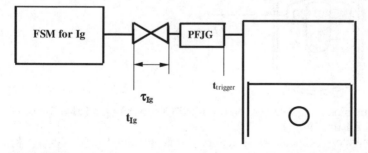

Fig. 3.6. Schematic diagram of a jet igniter module (Oppenheim et al. 1996)
PFJG: Pulsed Flame Jet Generator

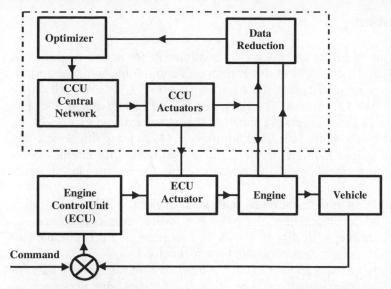

Fig. 3.7. Schematic diagram of a Combustion Control Unit (CCU) operating in conjunction with Engine Control Unit (ECU) (Oppenheim et al. 1996)

3.3 Prerequisite

Above all, to modernize internal combustion engines it is the realization of the essential role played by combustion that deserves particular care and attention. The sole purpose of combustion is, after all, to generate pressure that provides the force to push the piston. Consequently, the primary sensor of the effectiveness with which the exothermic process of combustion is executed in an engine cylinder, should be the pressure measurement. It is, therefore, pressure diagnostics that should be the primary monitor of engine performance – an obvious fact that, so far, did not receive the recognition it deserves. The background and principles of this technology, as well as representative examples of its implementation, are provided in Part 2.

To start with, pressure diagnostics should be employed to assess the performance of the engine adopted as the object of the R & D program for its modernization. Thereafter, it should serve as the monitor of any progress made in this program. Finally, it should be acquired as the primary feedback sensor in the control system for the eventual product: the MECC system.

3.4 Evolution

Recent advances in the technology of combustion for piston engines has been based on the concepts of direct injection and inhibition of the formation of flames by distributing the exothermic process, as exemplified by the advent of HCCI engines. The next step should be associated with implementation of the concepts presented in this chapter, i.e. (1) jet injection, exemplified by the pulsed air-blast atomizer, and (2) jet ignition executed by means of pulsed flame jet igniters, developed further for engines operating at high rotational speeds by cross-breeding them with plasma igniters.

To operate an engine equipped with such devices, a MECC system is necessary. The development of these devices has to be associated and coordinated, therefore, with that of MECC – a program of R&D commensurate in sophistication to that conducted for fuel cells.

Eventually this should lead to such a high level of technology that engine-out pollutant emissions could be of an order comparable in concentration to those of ambient air in a crowded city, while fuel consumption is cut down by, at least, a factor of two, as shown in the next section. The advantages accrued thereby should be equivalent to those expected of fuel cells. Consider, in particular, the SOFC[10], adopted recently by BMW for auxiliary power. In contrast to the PEMFC[11] operating at room temperature, but restricted to the use of hydrogen only, the SOFC allows the utilization of carbon monoxide, but operates at a temperature exceeding 1000 K. This is above the temperature of chain branching for oxidation of a hydrocarbon fuel in air – the fundamental threshold for ignition. Under such circumstances, in fact, combustion offers more favorable conditions for an intermittent operation, rather than those afforded by the steady flow system of a fuel cell.

3.5 Implementation

Today, the electronic computer and control technologies are so advanced, that, with their use, one should be able to test any contemplated technique for executing the exothermic process of combustion by an entirely computational diagnostics, before any hardware for this purpose is available.

A prognostic design procedure of this kind is presented here in Chap. 6. Starting with a given engine whose performance is to be improved by

---

[10] Solid Oxide Fuel Cell
[11] Proton Exchange Membrane Fuel Cell

means of a MECC system, its performance is first assessed by pressure diagnostics based on the dynamometer test data. Then, the consequence of introducing each step to control the execution of the exothermic process of combustion is examined. The first of them is the prospective gain expected from rendering a mixed charge system to operate, like an unmixed charge (diesel) engine, part load at wide open throttle. The second is to explore the improvements in fuel economy and reduced pollutant emissions due to protection from heart losses of the zone where the exothermic reaction takes place by keeping it away from the walls and by running it at higher speeds. The former can be executed by the use of a "smart head" featuring axially directed turbulent jets as illustrated by Fig. 3.2. The latter is made possible by the appreciably faster propagation mechanism provided by the fireball mode of executing the chemical reactions in parallel in contrast to the in series processing by flames traversing the charge.

The major purpose of such procedures for implementation of the MECC system is to inhibit the formation of discrete high temperature peaks, achievable at flame fronts, and prolong significantly the residence time of the reacting particles in the zones of the exothermic chemical reaction.

The results of such an assessment considering only the gains advantages of just distributing the combustion zone and protecting it from heat transfer losses in the course of the exothermic stage, carried out for a Renault engine, demonstrate that peak temperature can be thus reduced to a level at which the production of NO is practically annihilated, while, concomitantly, the concentration of CO is cut down by an order of magnitude, while fuel consumption for the same power output is halved. Concomitantly the yield of unburned hydrocarbons can be drastically reduced by proper control of the combustion field executed by turbulent jets, combined eventually with complete separation of the combustion zone in the cylinder and the lubrication zone in the crankcase achievable by relieving the piston from the demand to provide service as a crosshead for the crank mechanism in the crankcase, as illustrated by Fig. 3.3.

The method of approach on which this study is based and specific examples of its implementation are presented in Part 2.

PART 2

Analysis

4 Diagnosis

4.1 Introduction

As pointed out at the outset, the sole purpose of combustion in a piston engine is the generation of pressure in order to shift the expansion process away from compression and create a work-producing cycle. This transition is manifested by dynamic symptoms: the measured pressure profiler and the concomitant piston motion. Its action is referred to, therefore, as the dynamic stage of combustion. The generation of pressure is motivated by thermochemical transformation of reactants into products taking place in the course of a process referred to as the exothermic stage. Changes in the composition and thermodynamic states of the reacting substances (metamorphosis) taking place within this stage are deduced from the measured time profile of pressure – the procedure of pressure diagnostics. Treated thus is an inverse problem: deduction of information on an action from its recorded outcome. Its solution yields, in effect, a measure of the effectiveness of the exothermic process of combustion in a piston engine, akin to the account of an auditor on the operation of a business enterprise from the records of its transactions – a well established methodology in acquiring guidance for its improvement.

The concept of pressure diagnostics is, of course, well known. Among numerous publications, it is featured in textbooks on internal combustion engines, as well as in algorithms for its numerical implementation.

Of particular interest among the former is the book of John Heywood (1988), where methods for evaluating the thermodynamic properties of the working substance are described in Chaps. 3 and 4, and the thermodynamic approach to engine combustion is presented in Chap. 9. Among the latter, straightforward examples of numerical analysis based on pressure measurements are provided by Gordon Blair in his books on two-stroke (Blair 1996) and four-stroke (Blair 1999) engines.

The conventional treatment of the inverse problem concerned with the interpretation of measured pressure data is known as "heat release analysis". In relevance to piston engines, it stems from the classical paper of Rassweiler and Withrow (1938) and is usually restricted to two-component mixtures, referred to popularly as a two-zone model, disregarding the fact that only part of the charge participates in the transformation of reactants into products.

The method of approach presented here takes into account the multi-component composition of the working substance and bears, in particular, upon the exothermic stage of combustion with its effective and ineffective parts[1]. It is the minimization of the latter that is of key significance to progress in the technology of combustion in piston engines.

4.2 System

4.2.1 Component Mass Fractions

A system, in a thermodynamic sense, is matter, referred to as the working substance, enclosed within an impermeable boundary. The working substance is identified by its composition specified in terms of the component mass fractions, y_K and thermodynamic parameters of the component states, z_K. In the course of the exothermic stage, the cylinder charge undergoes a transformation (metamorphosis) from the reactants, y_R, to products, y_P, – a process involving evolution of component fractions and their states.

In the analysis of combustion in piston engines, the reactants are considered as fuel and air mixed on a scale at which they can get engaged in a chemical reaction that, in principle, requires their molecular proximity. Thus, if the components of the cylinder charge are not sufficiently well mixed, as is normally the case, their mass fraction, $Y_R = Y_F + Y_A < 1$. The mass fraction of the rest of the charge, Y_B, not engaged in chemical reaction, is made out of the recirculated residual or exhaust gas, as well as the unburned portion of the fresh charge due to incomplete combustion. In terms of the air/fuel ratio in the reactants, $\sigma_R \equiv Y_A/Y_F$, that, in principle, is different from σ_a - the air/fuel ratio of the fresh mixture admitted at inlet to the cylinder, the mass fraction of reactants in the cylinder charge is

$$Y_R = (1 + \sigma_R)Y_F \qquad (4.1)$$

Evolution of component mass fractions taking place in the course of the exothermic stage, when the reactants undergo a transformation (metamorphosis) into products, is described in Fig. 4.1 – a graphical representation of the mass balance for a fuel-lean system[2]. On its basis,

---

[1] Gavillet et al. 1993; Oppenheim et al. 1994; Oppenheim and Kuhl 1998, 2000; Oppenheim et al. 2000; Oppenheim et al. 2001; Shen et al. 2002; Shen et al. 2003

[2] For a fuel-rich system, identification of the components is reversed by interchanging subscripts F and A.

$$y_P(t) = Y_R x_F(t) \tag{4.2}$$

while, with (4.1) taken into account.

$$y_A(t) = Y_A (1 - x_F) = \frac{\sigma_R}{\sigma_R + 1} Y_R [1 - x_F(t)] \tag{4.3}$$

If, following common practice, the air/fuel ratio, σ_R is considered to be invariant, these functions are linear, as depicted on Fig. 4.1. Their ordered set forms a vector of components, **y**.

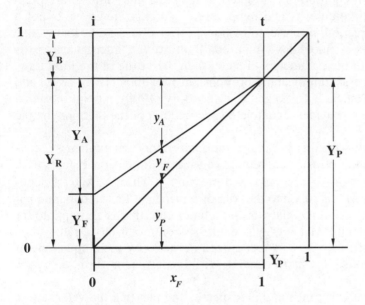

Fig. 4.1. A diagram of component mass fractions

In its role for identifying the composition and state of the reactants, the air/fuel ratio, σ_R, is of significance to the state of the products, in peticular their temperature. Its value is determined, in principle, by in *situ* measurement. In practice, it is inferred from chemical exhaust gas analysis, oxygen sensors, or laser diagnostics (vid. Warnatz et al. 1996). If the actual value of the air/fuel ratio in the reacting medium is not known, one can evaluate only the two limits within which it is bound: a completely unmixed system, and a perfectly mixed system. At the unmixed limit, the exothermic reaction is executed by flames, at which the air/fuel ratio is stoichiometric, so that the local air excess coefficient $\lambda = 1$, while the products reach adiabatic flame temperature - the maximum achievable by fuel irrespectively of the actual amount of air. At the mixed limit, the exothermic reaction is executed by a perfectly distributed combustion – the epitome of HCCI and MECC, where the air/fuel ratio is fixed by the composition of the cylinder charge that, for a desirable lean system, corresponds to $\lambda > 1$, and the products are formed at significantly lower temperatures.

4.2.2 Component States

The thermodynamic state of a component is identified by three parameters, if it is at its own equilibrium. Adopted usually for this purpose are pressure, p, specific volume, v, and temperature, T. This, however, is not a complete specification, because, to evaluate internal energy, e, knowledge of specific heats is, moreover, required. If, however, internal energy is adopted as one of the parameters of state, then, by virtue of the First Law, according to its essential meaning brought out by Gibbs (1875–1878) and formulated by Poincaré (1892) and Carathéodory (1909), a point in such a coordinate system provides a complete specification of the thermodynamic state.

In accordance with this principle, internal energy per unit mass, e, is adopted here as the primary coordinate of a three-dimensional state space in conjunction $w \equiv h - e = pv = p/\rho$, and pressure p. Thus, w, known popularly as "flow work" is cast into the role of a principal thermodynamic parameter of state, taking the place of the temperature that is appropriate for this purpose when internal energy is expressed on molar basis, rather than mass. The significance of w is recognized *de facto* by most equations of state, as manifested by their general, virial coefficient form, $w = pv = f(T,v)$.

Of particular significance in this respect is the fact that the values of w, as well as of e, are readily obtainable from thermodynamic tables, such as JANAF (Stull and Prophet, 1971) or NIST Chemistry WebBook[3], and calculable by the use of computational programs, such as STANJAN (Reynolds 1996), or CEA (Gordon and McBride 1994, McBride and Gordon 1996)[4].

The thermodynamic state of the system and of any of its components is identified by a state vector, $\mathbf{z}$, in a Cartesian three-dimensional phase (or state) space in terms of polar coordinates where $\mathbf{z}$ is the radius vector. The locus of states of a component is then *de facto* a vector hodograph – a trajectory in the three-dimensional thermodynamic phase space. Projections

---

[3] <http://webbook.nist.gov>
[4] <http://www.grc.nasa.gov/WWW/CEAWeb/>

of these trajectories on a plane of constant pressure, identified by their two coordinates, $z_K = w_K, e_K$, are displayed on the state diagram of Fig. 4.2.

The coordinates of reactants, R, are evaluated on the basis of their composition specified by the air/fuel ratio, σ_R, while for the products, P, they are determined by the condition of thermodynamic equilibrium that, for a single component, is specified, according to the Gibbs phase rule, by two parameters of state. Displayed on the diagram are state vectors of processes that take place in the simplest case of an adiabatic and isochoric combustion system (circles denoting the initial states and squares the terminal points).

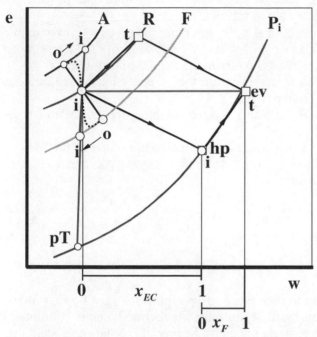

Fig. 4.2. A state diagram for an adiabatic and isochoric system on the plane of initial pressure, $p = p_i$

The plane of this diagram, corresponding to a fixed pressure, is visualized conveniently as a platform that is elevated by pressure – a displacement associated, in general, by distortion in the shapes of the state trajectories. This feature is of particular significance to the products, P, since, in order to comply with the condition of thermodynamic equilibrium, their composition and, hence, molar masses vary in the course of chemical reaction. The reactants, R, on the other hand, are usually treated as perfect gases, because, according to JANAF tables (Stull and Prophet 1971), their components, A and F, are treated as perfect gases. Consequently, their trajectories are the same at all pressure levels, so that, in the three-dimensional state space, they form a curved wall parallel to the pressure axis. If, moreover, they are considered as perfect gases with constant specific heats, this wall is plane and their projections at any pressure level collapse into a straight line.

A change of state that can be expressed by a transformation vector, z_{a-z} $(a - z = R - P, I - t)$ - a vector difference between polar vectors of the terminal state, t, and of the initial state, i. Such vectors delineating the four changes of state taking place at the bounds of an exothermic process are displayed in Fig. 4.2. They are as follows:

1. from i on R to i on P, both at p_i, representing the process of exothermic reaction from $x_{EC} = 0$ to $x_{EC} = 1$ where $x_{ES} = 0$, subscript EC denoting an exothermic center, while ES refers to the exothermic system.
2. from i on R to t on R, representing the compression of reactants as pressure increases from p_i to p_t.
3. from t on R to t on P, both at $p_t > p_i$, representing the exothermic reaction at $x_{ES} = 1$.
4. from i on P at p_i to t at p_t, representing the compression of products as pressure increases from p_i to p_t, while the progress parameter changes from $x_{ES} = 0$ to $x_{ES} = 1$.

4.3 Processes

4.3.1 Mixing

For a chemical reaction to take place, its components must be first mixed to form a molecular aggregate. If, initially, the thermodynamic coordinates of fuel and air are different, they have to be brought to the same state i on R – a task accomplished physically by transport processes of molecular mass diffusion and thermal conduction, assisted by viscosity. In Fig. 4.2, the concomitant changes of state taking place in the course of mixing are expressed by broken curves between points o on A and o on F to point i on R, with their directions indicated by arrows. The effect of mixing is manifested by rotation of the end point of the state vector around point i on R. Irrespectively of the influence of molecular diffusion, which, as a rule, must be involved in forming the reacting mixture, its outcome can be identified right from the outset by the intersection of the straight line between points o on A and o on F with R.

4.3.2 Exothermic Center

An exothermic center is a site of the exothermic reaction. It constitutes the kernel of the exothermic stage of combustion. Exothermic centers have been known for a long time in detonation literature as "hot spots". Their non-steady behavior under the influence of molecular diffusion has been studied extensively as the process of ignition (Boddington et al. 1971; Gray and Scott 1990; Griffiths 1990). Their diffusion dominated steady state model is a laminar flame. Their non-steady version in a turbulent field is referred to as the "flamelet model" (Peters 2000).

The fluid mechanical features of exothermic centers were investigated experimentally and theoretically in connection with their relevance to detonation and explosion phenomena, leading to the identification of mild and strong ignition centers (Oppenheim 1985). In a gasdynamic field where exothermic reaction takes place, exothermic centers occur at discrete sites. Each of them behaves then as a point singularity - a constant pressure deflagration center, where a finite change of state takes place locally at constant pressure - rather than across a straight line as it does in the classical version of a deflagration front.

4.4 Exothermic System

4.4.1 Evolution of Products

Changes of state taking place in the general case of an exothermic system are displayed in Fig. 4.3.

State coordinates of the system, $z_s = w_s$, e_s, or $z_s = w_s$, h_s, are given at any pressure level by a scalar product of the vector of components and the vector of states. Thus, with reference to Fig. 4.1,

$$z_S = \mathbf{y} \cdot \mathbf{z} = y_F z_F + y_A z_A + y_P z_P + y_B z_B \tag{4.4}$$

whence, in view of (4.1) and 4.2),

$$z_S = Y_F z_F + Y_A z_A + Y_B z_B + (Y_P z_P - Y_F z_F - Y_A z_A) x_F \tag{4.5}$$

The part of z_s independent of x_p identifies the vector of charge, expressed in terms of its components

$$z_C \equiv Y_F z_F + Y_A z_A + Y_B z_B \tag{4.6}$$

The thermodynamic parameters of this vector are variable, while the composition is fixed by initial conditions.

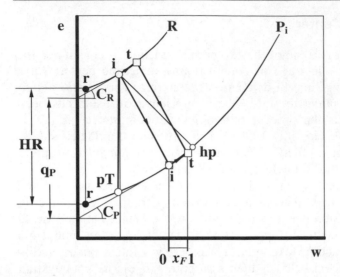

Fig. 4.3. A state diagram for an exothermic system

The reactants, R, are made out by mixing fuel, F, with air, A, as illustrated in Fig. 4.2. The state parameters of R are expressed in terms of mass averaged quantities

$$z_R = \frac{Y_F z_F + Y_A z_A}{Y_F + Y_A} = \frac{z_F + \sigma_R z_A}{1 + \sigma_R} \tag{4.7}$$

for which, according to (4.1), σ_R is usually prescribed.

Thus, in view of (4.2), (4.6) yields a fundamental expression for the exothermic center

$$z_S - z_C = (z_P - z_R) y_P \tag{4.8}$$

In a combustion field, an exothermic center is, as pointed out in the previous section, a discontinuity, where a jump from **i**, where $x_F = 0$, to **t**, where $x_F = 1$, takes place, while its pressure remains unchanged, while, according to (4.8) with (4.2), its amplitude is

$$z_{St} - z_{Si} = Y_R (z_P - z_R) \tag{4.9}$$

4.4.2 Coordinate Transformation

As indicated in Fig. 4.3, the ordinates of the polar of states, u_K, are transformed, for the sake of convenience, into component coordinates by the use of the gradients of state vectors, $C_K \equiv (e_t - e_i)_K / (w_t - w_i)_K$ (K = R, P).

The distance between them is expressed by the difference between intersections of their lines with the axis of ordinates, $q_p = e_{R_o} - e_{P_o}$ – referred to as the exothermic energy. Provided thereby is a strict, geometric, measure of a reference quantity that obviates the need for the conventional parameter of "heat release" (HR), whose use leads to misconceptions when it is treated as a constant in conjunction with fixed specific heats. Its awkward role with variable specific heats is evident in Fig. 4.3, where the reference states for it are denoted by symbol r.

Thus

$$e_R = C_R w_R \qquad (4.10)$$

while

$$e_P = C_P w_P - q_P \qquad (4.11)$$

Since the initial state of the system is *de facto* the initial state of the charge, similarly as for (4.10),

$$e_C - e_{Si} = C_C (w_C - w_{Si}) \qquad (4.12)$$

whereas, in view of (4.10) and (4.11),

$$e_R - e_P = C_R w_R - C_P w_P + q_P \qquad (4.13)$$

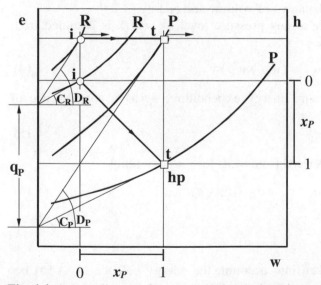

Fig. 4.4. A state diagram for an adiabatic exothermic center in *w-e* and *w-h* coordinate systems

As an illustration of this expression, a transformation from a state of reactants, i, to that of products, f, for an adiabatic exothermic center, evaluated according to it, is provided by the state diagram of Fig. 4.4. Displayed here also is the equivalent expression for enthalpy, $h \equiv e + u$ - the sum of its coordinates - adopted as the ordinate instead of internal energy. According to it, a change of state taking place in the course of an adiabatic exothermic reaction proceeds along a constant enthalpy path - a diagonal on the e-w plane, expressed by $e = h_i - w$, or just a change of the reference parameter, Δw, on the h-w plane. As implied by the geometry of the diagram, the slope $D_K = C_K + 1$ ($K = R, P$). If the reactants are considered as perfect gases with constant specific heats, $c_{Kv} \equiv (\partial e_K/\partial T)_v$ and $c_{Kp} \equiv (\partial h_K/\partial T)_p$, then, in terms of $R_K \equiv R/M_K$, where R is the gas constant and M_K the molar mass, while $\gamma_K \equiv c_{Kp}/c_{Kv}$, the two slopes are $C_K = c_{Kv}/R_K = 1/(\gamma - 1)$ and $D_K = c_{Kp}/R_K = \gamma/(\gamma - 1)$.

4.5 Closed System

4.5.1 Balances

By definition, a thermodynamic system is confined within an impermeable boundary. The effects of the exothermic reaction it undergoes are expressed by (4.8). Changes of states taking place thereby in a closed system are determined by the balances of volume and energy.

The volume balance at any pressure level, $p_K = p(t)$, is obtained from (4.8) for $z = w$, whence

$$w_S - w_C = (w_P - w_R)\, y_P \tag{4.14}$$

Then, taking into account energy expenditure, e_e, as a consequence of which

$$e_S = e_{Si} - e_e \tag{4.15}$$

the energy balance is derived from (4.8) for $z = e$, yielding

$$(e_P - e_R)\, y_P = e_{Si} - e_C - e_e \tag{4.16}$$

4.5.2 Products

With (4.10)–(4.13) taken into account, the energy balance of (4.16) becomes

$$(C_R w_R - C_P w_P + q_P)\, y_P = C_C (w_C - w_{Si}) + e_e \tag{4.17}$$

whence, upon eliminating w_p by the use of volume balance expressed by (4.14), the mass fraction of products

$$y_P = \frac{C_P(w_S - w_C) + C_C(w_C - w_{Si}) + e_e}{q_P - (C_P - C_R)w_R} \tag{4.18}$$

In the case of a closed system, the energy expenditure

$$e_e \equiv w_w + e_I \tag{4.19}$$

where, in terms of v_S denoting the ratio of cylinder to clearance volume, v_C, per unit mass of the working substance, $w_w \equiv v_c \int_{v_{Si}}^{v_S} (p - p_b)dv$ is the work performed by the system, while e_I, expresses the energy undetectable by the measured pressure profile. In the former, p_b denotes the backpressure, be it variable or constant, including a judicious increment, if one wishes to account for the effects of friction. The latter is made out, primarily, of energy loss incurred by heat transfer to the walls - the major irreversibility of the exothermic stage – with a small part that may be due to leakage.

4.6 Exothermic Stage

The solution of the inverse problem expressed by (4.18), can be cast into an explicit function of pressure by
(1)   normalizing its variables with respect to the initial state, **i**, and
(2)   considering the compression of the unreacted medium, taking place in the course of the exothermic stage, as polytropic with the same index, n, as that of the preceding it process of compression by the piston.

Thus, in terms of $W_K \equiv w_K / w_{Si}$ (K = S, R, P), while by definition of w_K, $W_S = PV_S$, where $P \equiv p/p_i$, $V_S \equiv v_S/v_{Si}$, whereas $W_e \equiv e/w_{Si}$ and $Q_P \equiv q_P / w_{Si}$, it follows from (4.18) that

$$y_P = \frac{C_P(W_S - W_C) + C_C(W_C - W_{Si}) + W_e}{Q_P - (C_P - C_R)W_R} \tag{4.20}$$

while

$$W_R = P^m \tag{4.21}$$

where $m \equiv 1 - n^{-1}$.

Concomitantly, from the volume balance of (4.14) it follows that

$$W_P = W_R + \frac{W_S - W_C}{y_P} = P^m + \frac{PV - P^m}{y_P} \qquad (4.22)$$

The mass fraction of the generated products is considered to consist of an effective part, y_E, and ineffective part, y_I, i.e., taking into account (4.2),

$$y_P = Y_R x_F = y_E + y_I \qquad (4.23)$$

where, according to (4.19) and (4.20), in terms of $W_w \equiv w_w / w_{Si}$,

$$y_E = \frac{C_P(PV - 1) - (C_P - C_C)(P^m - 1) + W_w}{Q_P - (C_P - C_R) P^m} \qquad (4.24)$$

while, with $e_I / w_{Si} \equiv Q_I$,

$$y_I = \frac{Q_I}{Q_P - (C_P - C_R) P^m} \qquad (4.25)$$

The exothermic stage is defined as the process in the course of which the effective mass fraction of generated products, y_E, reaches its maximum level, identifying thus the terminal state, **t**, as expounded in Sect. 4.9. It starts at the initial state **i** where, by definition $y_E = 0$. Since this state is the essential singularity of combustion, as brought up in Sect. 4.9, its identification has to be deduced from the evolution of the polytropic pressure model in the course of the dynamic stage – a concept identified in Sect. 4.11.

It is of interest to note that, if $C_P = C_C = C_R$, (4.24) is reduced to

$$y_E = \frac{(PV - 1) + W_w}{Y_R Q_P / C_P} \qquad (4.26)$$

– the key to what is known as "heat release analysis" – a well established concept in engine performance analysis, on the basis of the classical paper of Rassweiler and Withrow (1938) and, in combustion literature, by the classical text of Lewis and von Elbe (1987). In an isochoric system, when $V = 1$ and $W_w = 0$, under adiabatic condition of $Q_I = 0$, when exergy is at its maximum, obtained thereby is the popular proportionality relationship between the mass fraction of the generated products, y_P, and the measured overpressure, $P - 1$.

Its modern version is expressed popularly by the concept of the so-called "apparent heat release",

$$Q_{app} \equiv q_{app}\, y_E = \frac{C_P w_{Si}}{Y_R} \frac{(PV-1)+W_w}{P_t V_t -1 + W_{wt}} \qquad (4.27)$$

Provided thus is also the key to what is known as a "zero dimensional analysis". In the case of $Y_P = 1$, it is referred to as one "zone", while for $Y_P < 1$ it is considered as two 'zones' or more, depending on the number of components one wishes to take into account.

Implementation of pressure diagnostics provides specific data on the pragmatic features of combustion, among which the rate of fuel consumption, referred to conventionally as the rate of burn, is of particular significance. To reveal its evolution, including the profiles of the thermodynamic parameters of components of the exothermic system, the profile of mass fraction of the generated products, $y_P(\Theta)$, has to be expressed in form of sufficiently smooth (differentiable) analytic function. To accomplish that, one is faced with three questions:

(1) How to evaluate the mass fraction of consumed fuel, x_F, upon the determination, by means of (4.24), of the effective part of generated products, y_E?

(2) How to extract the signal from the measured engine test data that, perforce, include noise and, in particular,

(3) How to pinpoint the initial state, **i**, and the final state, **t**, of the exothermic stage, specifying the bounds within which the transformation of reactants into products takes place?

Answers to these questions are provided, respectively, by the next three sections.

4.7 Correlation

In order to coordinate the mass fraction of consumed reactants, x_F, with the effective part of g the generated products, y_E, evaluated by the use of (4.24), a study of heat transfer from combustion in a vessel of constant volume, equipped with thin film heat transfer probes and pressure transducers (Oppenheim and Maxson 1994, Oppenheim and Kuhl 2000a, 2000b) was conducted. Its highlights are recounted in Appendix A.

The ineffective energy expenditure, $e_r(t)$ was then identified with energy loss, $q_w(t)$, deduced from measured heat fluxes. Since, by virtue of (4.2) and (4.22), $y_{Et} + y_{It} = Y_R$, the mass fraction of consumed fuel, $x_F(t)$, as well as of its effective and ineffective parts, $x_E(t)$ and $x_i(t)$, were thereby determined.

On this basis it was established that, irrespectively whether combustion was propagated in a practically laminar or turbulent manner, the relationship between the normalized effective mass fractions of the consumed reactants, can be expressed in terms of the power function

$$x_E / x_{Et} = y_E / y_{Et} = 1 - (1 - x_F)^{\sigma / x_{Et}} \qquad (4.28)$$

displayed in Fig. 4.5, where, according to results of the heat transfer study recounted in Appendix A, it has been established that $\sigma = x_{Et}^{1/2}$.

A correlation function between the mass fraction of fuel consumed in the course of the exothermic stage, and the normalized effective part of the generated products, is given by the inverse of (4.28)

$$\boxed{x_F = 1 - (1 - \tilde{y}_E)^{\kappa}} \qquad (4.29)$$

where $\tilde{y}_E \equiv y_E / y_{Et}$ and $\kappa = x_{Et}^{1/2}$.

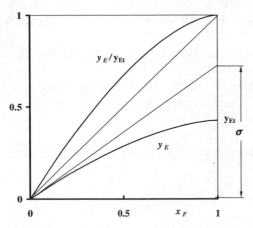

Fig. 4.5. Correlation between the total mass fraction of consumed fuel and the normalized effective part the mass fraction of generated products

The validity of the correlation for the case of variable cylinder volume in the course of the exothermic stage is based on the postulate that it cannot be affected by the relatively small deformation due to piston motion. Under such circumstances, the correlation function expressed by (4.29) provides means for evaluating the mass fraction of consumed fuel, $x_F(\Theta)$, directly from the normalized effective part of the generated products, $y_E(\Theta)$, established by (4.24). The results thus obtained were found to be in a satisfactory agreement with the well-known Woschni correlation[5] (Czerwinski and Spektor 2000).

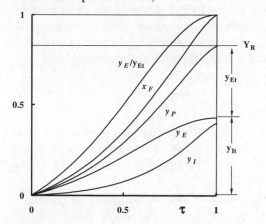

Fig. 4.6. Time profiles of mass fractions

4.8 Evolution

4.8.1 Initial State

The initial state of the exothermic stage of combustion, **i**, is an essential singularity of combustion, because the specific volume of products at this state, $v_i = V_i/M_i = 0/0$ (!) – the *raison d'être* of what is known in combustion literature as the "cold-boundary difficulty" for laminar flames (Williams 1985). Here this feature is manifested for turbulent combustion systems by (4.22) for $P = V = 1$ at $y_p = 0$.

Thus, depending on the round-off error, calculations of specific volume and temperature at the initial state of the products tend to end either at those of the reactants or at infinity. In effect, then, state **i** is located at a saddle point singularity specified by the intersection of its axes, a separatrix expressing the process of compression corresponding to $\pi_c = $ const. and an attractor prescribed by $y_p(\Theta)$, at a finite angle, forming thus a corner. Since nature abhors corners, the initial point, **i**, is obviated by experimental data.

However, this state is of pivotal significance to the dynamic and exothermic stages of combustion. To identify its coordinates, the process of ignition has to be taken out of scope. Some experimental data (usually quite small in number) in its immediate vicinity must be, therefore, disre-

---

[5] Woschni 1966/67, 1967, 1970, Woschni and Anisits 1973, 1974

garded, as prescribed by the Cauchy principle. The analysis treats then the dynamics of the exothermic process, irrespectively of the mechanism by which it is initiated, or of the form in which it takes place. Being, therefore, independent of the type in which the process of combustion is ignited or of the kind of flame by which it is executed, it is applicable to any flame or mode of combustion, in particular to its flameless, distributed kind.

4.8.2 Terminal State

The terminal state, $y_{P_t}(\Theta)$ is also a singularity because it is at the maximum marked by the balance between the positive effect of pressure generated in by the exothermic reaction and the negative influence of energy loss incurred by heat transfer to the walls.

Thus, in contrast to the sharp corner of the saddle point singularity at the initial state, **i**, the terminal state, **t**, is a nodal point singularity approached smoothly to the maximum of the effective mass fraction of the generated pro ducts, $y_E(\Theta)$, by which it is identified.

4.8.3 Transition

It is thus evident that, in order to obtain an expression for an integral curve satisfying the requirements of the two singularities – the basis for evaluating the dynamic properties of combustion expressed principally by its burning rate, a differential of the progress parameter – the evolution of the exothermic process has to be expressed in terms of a sufficiently smooth analytic function. This is accomplished by the life function presented in the next section.

4.9 Life Function

The essential purpose of pressure diagnostics is to express the behavior of the system as a dynamic object – the only kind that a control system can affect. Such an interpretation of the generation of products and concomitant consumption of fuel is provided by a progress parameter, $x(\tau)$, akin to displacement or distance of travel, with its slope akin to velocity and the change of slope akin to acceleration – a monotonic, smooth function of time. Between the singularities at its bounds, the initial point, **i**, and the terminal point, **t**, the trajectory expressed by $x(\tau)$ provides an analytic interpretation of the measured pressure data.

The progress parameter, $x(\tau)$, models, therefore, the evolution of life: it starts at a finite slope – the condition of birth – and ends at zero slope – the condition of death. For the exothermic process of combustion, it provides an analytic expression for its principal features: history of the evolution of fuel consumption (the rate of burn) and generation of products. It has to be, therefore, sufficiently smooth to be differentiable.

The rate of change, $\dot{x} \equiv dx/d\tau$, satisfying these conditions is expressed, in terms of the progress parameter for time is $\tau \equiv (t - t)/T$ where $T \equiv t_t - t_o$ is the lifetime of the exothermic stage, by the bi-parametric function

$$\dot{x} = \alpha\,(\xi + x)(1 - \tau)^{\chi} \qquad (4.30)$$

whence, at $\tau = 0$, $\dot{x}_i = \alpha\xi > 0$, and at $\tau = 1$, $\dot{x}_f = 0$, as required. By quadrature, subject to boundary conditions of $x = 0$ at $\tau = 0$ and $x = 1$ at $\tau = 1$, (4.30) yields the life function

$$x = \frac{e^{\zeta} - 1}{e^{\zeta_f} - 1} \qquad (4.31)$$

where

$$\zeta = \frac{\alpha}{\chi + 1}[1 - (1 - \tau)^{\chi + 1}] \qquad (4.32)$$

– an exponential of the same form as (4.28). It follows then that $\zeta_f = \dfrac{\alpha}{\chi + 1}$ and $\xi = \dfrac{1}{e^{\zeta_f} - 1}$. Profiles of representative life functions and of their derivatives are displayed in Fig. 4.5.

In view of (4.30), the change of rate

$$\ddot{x} \equiv d^2x / d\tau^2 = (\xi + x)(\dot{\zeta}^2 + \ddot{\zeta}) \qquad (4.33)$$

while, according to (4.32),

$$\dot{\zeta} = \alpha(1 - \tau)^{\chi} \qquad (4.34)$$

whence

$$\ddot{\zeta} = -\alpha\chi(1 - \tau)^{\chi - 1} = -\frac{\chi}{1 - \tau}\dot{\zeta} \qquad (4.35)$$

At the point of inflection (menopause), where $\ddot{x} = 0$, according to (4.33), with (4.34) and (4.35),

$$\alpha(1 - \tau^*)^{\chi + 1} - \chi = 0 \qquad (4.36)$$

whence

$$\tau^* = 1 - (\frac{\chi}{\alpha})^{\frac{1}{\chi + 1}} \quad \text{and} \quad \zeta^* = \frac{\alpha - \chi}{\chi + 1} \qquad (4.37)$$

On this basis, two types of combustion events can be identified: those for which $\chi < \alpha$, so that $\tau^* > 0$, and those of $\alpha < \chi$, when $\tau^* < 0$. The former were formed by mild ignition, producing flame kernels, while the latter were started by strong ignition, characteristic of distributed combustion. If $\alpha = \chi$, then $\tau^* = 0$, and the point of inflection is at the initial state, **i**.

The concept of life function was derived from an extensive literature background (Oppenheim and Kuhl 1998) recounted in appendix A. It is, in effect, a reverse of the Vibe function, known in the literature as the Wiebe function – a misnomer promulgated by the German translation of the original paper of Vibe (1956), promoted by Jante (1960) who, upon acquainting himself with it, pronounced it as the "combustion law," before Vibe's book (1970) was published. To demonstrate its reverse nature with respect to life function, Vibe functions are depicted, with profiles of their derivative, on Fig. 4.7, for confrontation with life functions displayed on Fig. 4.8 with the same parameters α and χ.

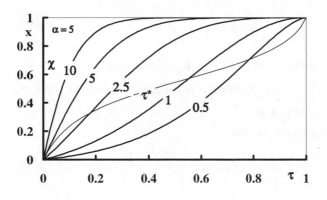

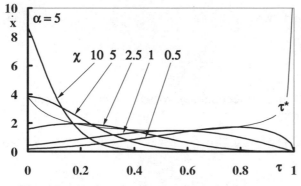

Fig. 4.7. Life functions and their derivatives

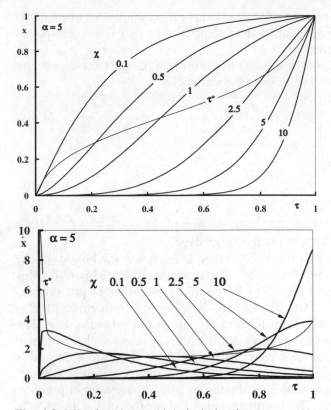

Fig. 4.8. Vibe functions and their derivatives

4.11 Dynamic Stage

To extract the signal from the noise incurred *per force* in the measured engine pressure data pertaining to the exothermic stage, they are interpreted in terms of a dynamic stage. In an enclosure of constant volume, this stage is manifested by the profile of a monotonic pressure rise from the initial state, **i**, to the final state, **f**, at its maximum. The location of neither of them is known *a priori*, for their positions are identified by the signal, rather than the intrinsically noisy data. A methodology for pinpointing the coordinates of these states is presented here.

In an enclosure of variable volume, such as the cylinder in a piston engine, this stage is described by analogy to the case of invariant volume, in terms of the polytropic pressure model

$$\pi \equiv p v_S^n \tag{4.38}$$

where v_s denotes the volume of the cylinder-piston enclosure normalized with respect to clearance volume, and the polytropic index, n, corresponds to the process of compression.

The dynamic stage is identified by the profile of the polytropic pressure model expressed in terms of its progress parameter

$$x_\pi(\tau_\pi) \equiv \frac{\pi(\tau_\pi) - \pi_i}{\pi_f - \pi_i} \qquad (4.39)$$

where

$$\tau_\pi \equiv \frac{\Theta - \Theta_i}{T_\pi} \qquad (4.40)$$

is the progress parameter for time as a function of the crank angle, Θ while $T_\cdot \equiv \Theta_f - \Theta_i$ is the lifetime of the dynamic stage.

In view of their singular nature described in Sect. 4.9, the bounds of the dynamic stage have to be pinpointed. The procedure adopted here for this purpose is described by Figs. 4.9–4.12 – typical examples of pressure diagnostics (Shen et al. 2001; Shen et al. 2002) whose implementation is presented in Chap. 5. The first displays the measured pressure data. The second shows the conventional indicator diagram in linear scales, obtained with given kinematics of the crankshaft mechanism. The third depicts this diagram in logarithmic scales, demonstrating how the polytropic index, n, for the process of compression is evaluated by linear regression of pressure measured in its course. The fourth presents the profile of the polytropic pressure model, π, defined by (4.38), obtained with its use.

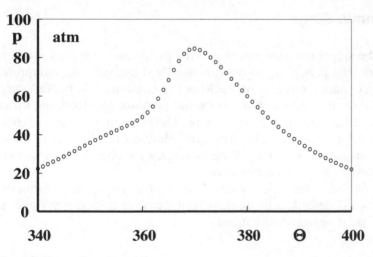

Fig. 4.9. Data of measured pressures

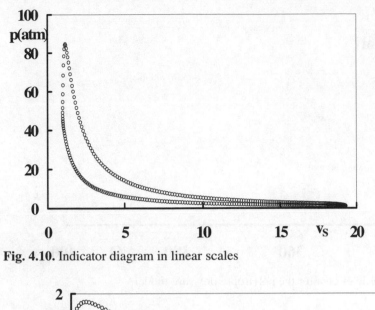

Fig. 4.10. Indicator diagram in linear scales

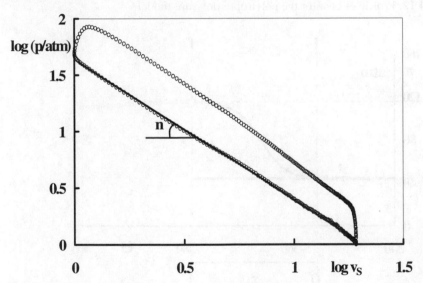

Fig. 4.11. Indicator diagram in logarithmic scales

The latter is expressed in an analytic form by regression to the life function introduced in the previous section - an iterative procedure establishing the unknown *a priori* locations of the initial state, **i** – the essential singularity of combustion – and the final state, **f**, at the maximum of $\pi(\Theta)$ that, in principle, is different from the terminal point of the exothermic stage, **t**, occurring, as a rule, somewhat later.

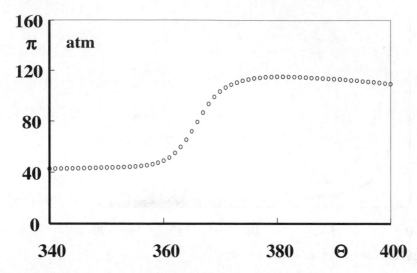

Fig. 4.12. Profile of data for the polytropic pressure model

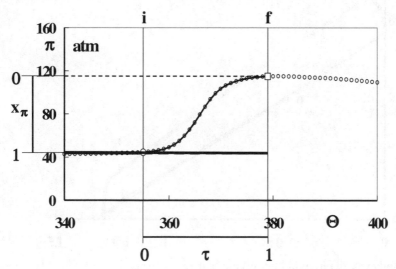

Fig. 4.13. Analytic interpretation of pressure data

An analytic interpretation of the recorded polytropic pressure profile in terms of the life function is displayed in Fig. 4.13, whereas Fig. 4.14 demonstrates the remarkable accuracy with which the measured pressure data are thus expressed – a universal outcome of their regression to this function encountered in all the cases of its application exemplified in Chap. 5.

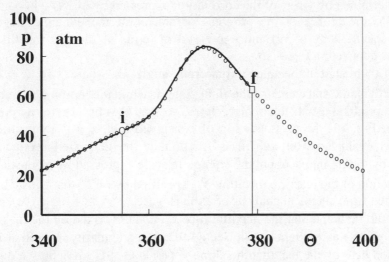

Fig. 4.14. Analytic interpretation of the pressure profile in comparison to its measured data

4.12 Thermodynamic Transformation

4.12.1 Progress Parameters

The profile of the normalized effective mass fraction of products generated in the course of the exothermic stage, $\tilde{y}_E(\Theta)$, evaluated on the basis of (4.24), is displayed in Fig. 4.15, together with that of the mass fraction consumed reactants, $x_F(\Theta)$ – the progress parameter of the exothermic

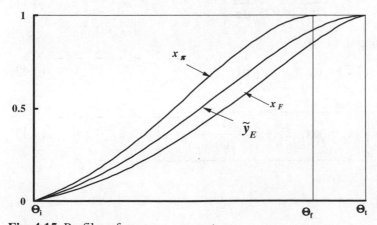

Fig. 4.15. Profiles of progress parameters

stage determined by virtue of the correlation expressed by (4.29) – in comparison to the profile of the progress parameter of the dynamic stage, $x_\pi(\Theta)$, specified by (4.39) and expressed in terms of the life function, (4.31), as depicted in Fig. 4.13.

All of them start at the same initial crank angle, Θ_t, where $\tilde{y}_E = x_F = x_\pi$ = 0. The dynamic stage ends at the final state, **f**, while the terminal state of the exothermic stage, **t**, takes place later, so that $\Theta_t \geq Q_t$, where, as presented in Fig. 4.1, the mass fraction of consumed fuel, $x_r = 1$. Its profile, $x_F(\Theta)$, is evaluated from the mass fraction of the generated products, $x_P(\Theta)$, by taking into account the change in scale expressed by a constant mass fraction of the reacting medium, Y_R, specified by (4.2), that is usually determined from chemical analysis of exhaust gases.

It should be noted that the maximum of $x_P(\Theta)$ at Θ_t, fixed by the maximum of $y_E(\Theta)$ as brought out in Sect. 4.9.2, is a singularity. Its effect is manifested here by the fact that the slope $x_F(\Theta)$ at Θ_t, $[dx_F/d\Theta]_t > 0$ – a discontinuity, as demonstrated in Sect. 4.12.2.

4.12.2 Rate of Fuel Consumption

Pragmatically, the most significant trait of combustion is the rate in which fuel is consumed, the rate of burn. The concepts of the correlation and the life function provide means for its analytic expression.

Thus, upon expressing the normalized effective mass fraction of consumed fuel, $\tilde{y}_E(t)$, evaluated by means of (4.24), in terms of a life function, by regression yielding its parameters α_E and χ_E, it follows from the correlation of (4.29) that the rate of fuel consumption

$$\dot{x}_F = \frac{\kappa}{(1-\tilde{y}_E)^{1-\kappa}} \frac{d\tilde{y}_E}{d\tau} \tag{4.41}$$

while according to (4.30),

$$\dot{\tilde{y}}_E = \alpha_E(\xi_E + \tilde{y}_E)(1-\tau)^{\chi_E} \tag{4.42}$$

whence

$$\boxed{\dot{x}_F = \kappa\alpha_E(\xi_E + \tilde{y}_E)\frac{(1-\tau)^{\chi_E}}{(1-\tilde{y}_E)^{1-\kappa}}} \tag{4.43}$$

According to the above, at $\tau = 0$ where $\tilde{y}_E = 0$, $\dfrac{dx_F}{d\tau} = \kappa\alpha_E\xi_E > 0$,

whereas at $\tau = 1$ where $\tilde{y}_E = 1$, $\dfrac{dx_F}{d\tau} = \dfrac{0}{0}$ - a singularity (!).

Thereupon, the rate of change in the rate of fuel consumption, the kinematic parameter akin to acceleration,

$$\ddot{x}_F = [k\alpha_E \frac{A - \tilde{y}_E}{1 - \tilde{y}_E}(1 - \tau)^{\chi_E} - \frac{\chi_E}{1 - \tau}]\dot{x}_F \qquad (4.44)$$

4.12.3 Thermodynamic Parameters

The corresponding temperatures and densities are evaluated on the basis of (4.22), in conjunction with the equation of state, according to which the temperature of the products $T_P = R_{Ri}T_{Ri}(W_P/R_P)$, where R_K (K = R, P), are the gas constants. Their profiles are depicted by Figs. 4.16 and 4.17. Consequences of singularities at both boundaries of the exothermic stage are manifested there distinctly by jumps from R to P at Θ_i and Θ_f.

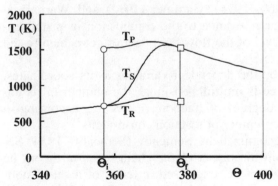

Fig. 4.16. Temperature profiles

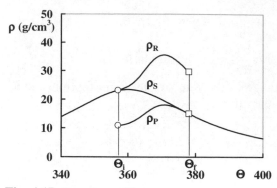

Fig. 4.17. Density profiles

4.13 Chemical Transformation

4.13.1 Background

Chemical transformations of reactants, R, into products, P, are of essential significance to the dynamics of exothermic systems. On the microscopic level of molecular interactions it is treated by chemical dynamics (vid. e.g. Steinfeld et al. (1989, 1999)) and chemical kinetics – its statistical interpretation (vid. e.g. Benson (1960, 1976), Gardiner (1972) and Warnatz al (1996), the latter of particular relevance to the combustion in piston engines). On the macroscopic level of the flow field their effect is manifested by exothermic centers.

A phase space is formed by the dependent variables as its coordinates. A thermochemical phase space is multidimensional, the number of its coordinates being equal to the degrees of freedom, $F = N + 2$, according to the Gibbs rule, where N is the number of reaction constituents.

This has been, in fact, recognized by Semenov (Semenoff 1934, Semenov 1958–1959) right from the outset of the chain reaction theory he founded. Its mechanism is, therefore, expressed in terms of a set of non-linear ordinary differential equations (ODEs), autonomous with respect to time - the sole independent variable of an autonomous set of ODEs[6] that, as typical of non-linear dynamics, is eliminated by a simple expedient of dividing the constitutive equations by each other.

Formulated thus are the fundamental differential equations describing the chemical kinetic behavior of the system. Their solutions, subject to appropriate constraints and initial conditions, are expressed on the thermophysical phase space in terms of integral curves, or trajectories. Their

---

[6] Ordinary Differential Equations

properties are most conveniently exhibited by their projections on the planes of any two co-ordinates one wishes to inspect.

In particular, the projection of the trajectories of the thermochemical phase space upon the plane of the temperature and the concentration of a chain carrier played an important role in the physical chemistry of combustion. It is on its basis that B.F. Gray and C.H. Yang (Gray and Yang 1965; Yang and Gray 1967) developed a concept of the "unification of the thermal and chain theories of explosion limits." They demonstrated that the effect of chain branching, referred to as the "chemical kinetic explosion", is manifested by a saddle-point singularity located at the intersection of a separator between the trajectories of reactants and the attractor of the products. Thereupon, P. Gray and his collaborators expanded this method of approach with remarkable success to a number of chemical systems (Gray and Scott 1990). Particularly noteworthy is the elucidation, provided by Griffiths (1990), of the role played by singularities in the course of thermo-kinetic interactions, encompassing a variety of nodal and saddle, as well as spiral points, the latter expressing the oscillatory behavior of cool flames.

At later stages of the exothermic process of combustion, the integral curves tend to bunch together, forming manifolds attracted by the co-ordinates of thermodynamic equilibrium, as demonstrated by Maas and Pope (1992a, 1992b). Provided thus is evidence that, eventually, the reacting system gets under the influence of the final state of thermodynamic equilibrium to such extent that, in its asymptotic approach, it becomes virtually independent of its initial stages of ignition and chain branching.

In order to reveal the dynamic features of an exothermic center, the influence of molecular diffusion, which tends to obscure the chemical kinetic mechanism, is, for the sake of clarity, taken out of scope, while the effect of thermal conductivity is expressed in terms of a relaxation time (Oppenheim 1985). The evolution of chemical reaction is thus considered as a transformation from the state of reactants into that of products. The former is at given composition and temperature, $T_R = T_i$, and the latter is at the terminal temperature, T_t, specified either by the thermodynamic equilibrium of products, so that $T_t = T_p$, or by the surroundings, in which case $T_t = T_s$.

4.13.2 Procedure

The equations of chemical kinetics pertain to the chemical source and to the thermal source.

The chemical source is formulated in terms of the so-called law of mass action, which, in the form of the popular computer algorithm, CHEMKIN

(Kee et al. 1980, Kee et al. 1989, Kee et al. 1993) is expressed by the species conservation equation in the following manner.

For K chemical reactant species, A_k ($k = 1, 2, ... K$), reacting in I elementary steps ($i = 1, 2, ... I$), each of the form

$$\sum_{k=1}^{K} a'_{ki} A_k \Leftrightarrow \sum_{k=1}^{K} a''_{ki} A_k \qquad (4.45)$$

where a'_{ki} and a''_{ki} denote stoichiometric coefficients of, respectively, the reactants and the products, the rate of gain in the mass fraction of the kth component

$$\frac{dy_k}{dt} = v_s M_k \sum_{i=1}^{I} (a''_{ki} - a'_{ki}) \{ k_i^+ \prod_{i=1}^{I} [A_k]^{a'_{ki}} - k_i^- \prod_{i=1}^{I} [A_k]^{a''_{ki}} \} \qquad (4.46)$$

for which the reaction rate constants

$$k_i = A_i T^{n_i} \exp(-E_i / RT) \qquad (4.47)$$

while v_s is the specific volume of the system and M_k is the molar mass of the kth component.

The thermal source is expressed in terms of the energy balance for an exothermic center

$$dq = de + pdv = dh - vdp \qquad (4.48)$$

where

$$de = \sum_{k=1}^{K} y_k de_k + \sum_{k=1}^{K} e_k dy_k \qquad (4.49a)$$

or

$$dh = \sum_{k=1}^{K} y_k dh_k + \sum_{k=1}^{K} h_k dy_k \qquad (4.49b)$$

Then, as a consequence of the fact that, according to the JANAF data base, the reaction constituents behave as perfect gases,

$$de_k = c_{v,k} dT \qquad (4.50a)$$

or

$$dh_k = c_{p,k} dT \qquad (4.50b)$$

so that the rate of the temperature rise, manifesting the action of the thermal source, can be expressed as

$$\frac{dT}{dt} = -\frac{1}{c_v}\sum_{k=1}^{K}e_k\frac{dy_k}{dt} - \frac{p}{c_v}\frac{dv}{dt} + \frac{1}{c_v}\frac{dq}{dt} \tag{4.51a}$$

where, $c_v \equiv \sum_{k=1}^{K} c_{v,k}y_k$, or as

$$\boxed{\frac{dT}{dt} = -\frac{1}{c_p}\sum_{k=1}^{K}h_k\frac{dy_k}{dt} + \frac{v}{c_p}\frac{dp}{dt} + \frac{1}{c_p}\frac{dq}{dt}} \tag{4.51b}$$

where, $c_p \equiv \sum_{k=1}^{K} c_{p,k}y_k$.

The last term in these equations is due to thermal diffusion, involving a gradient in the flow field, expressed by a partial derivative. Its ordinary differential model, in which it is cast here, is expressed in terms of a thermal relaxation time, τ, so that

$$\boxed{\frac{dq}{dt} = \frac{T_t - T}{\tau_T}} \tag{4.52}$$

where τ_T is a prescribed constant – an assumption justified by the relative insensitivity of the results on its value (Oppenheim 1985).

Evolution of the chemical kinetic mechanism is then revealed by integrating (4.46) and (4.51) with (4.52), for a set of given initial conditions, specified in terms of the thermodynamic state parameters and concentrations of the chemical components of the system at state **i**. By making sure that $k^+/k^- = K$, the equilibrium constant, while T_f is specified by the analysis of the exothermic stage of combustion, the integration is, in effect, a solution of a double boundary value problem where the end state is that of equilibrium.

The solution is associated with an essential difficulty due to the fact that these equations are essentially stiff, as pointed out originally by Hirschfelder et al. (1964), imposing a demand for careful treatment to assure convergence. Such procedures are nowadays aided significantly by the CHEMKIN method of Kee (Kee et al. 1980; Kee et al. 1989, 1993)[7], in conjunction with chemical kinetic data provided by the LLNL Combustion Chemistry Group under the direction of C.K. Westbrook (Westbrook and

---

[7] <www.ca.sandia.gov/chemkin>

Pitz 1984; Curran et al. 1998)[8], whose web site, incidentally, refers to the HCT program for chemical kinetic calculations, rather than to CHEMKIN. Both programs deal with the problem posed by stiff equations and employ for that reason the LSODE (Linear Solver of Ordinary Equations) procedure of Hindmarsh (1971), using the multi-step integration method of Gear (1971). Prominent features of this type of numerical methods were reviewed by Bui et al. (1984).

---

[8] <www-cms.llnl.gov/combustion/combustion_home.html>

5 Procedure

5.1 Prescription

A step-by-step sequence of operations to be carried out to implement the technique of pressure diagnostics, presented in the previous chapter, is as follows.

1. From pressure data recorded with respect to crank angle, $p(\Theta)$, (vid. Fig. 4.9), using given kinematics of the crankshaft mechanism, expressed by the profile of the cylinder-piston volume normalized with respect to clearance volume, $v_s(\Theta)$, an indicator diagram, expressing the functional relationship, $p(v_s)$, is deduced (vid. Fig. 4.10).

2. The slope of compression polytrope, n, is evaluated by linear regression, as displayed by the indicator diagram in terms of logarithmic coordinates, Fig. 4.11. The measured pressure data are, thereupon, mapped by the polytropic pressure model, $\pi(\Theta)$, prescribed by (4.38) (vid. Fig. 4.12).

3. The progress parameter for the dynamic stage, $x_.(\tau_\pi)$, specified by (4.38), is expressed then in analytic form by regression to a life function defined by (4.31), yielding its parameters, λ_π and χ_π Fixed thereby are (1) the final state of the dynamic stage, **f**, at the maximum of $\pi(\Theta)$, where , $x_\pi(\tau_\pi) = 1$, and (2) the initial state of the dynamic stage, **i**, located at the intersection of $\pi(\Theta)$ with the base line, while, according to the Cauchy principle pointed out in Sect. 4.8.1, data points in its immediate vicinity are discarded.

4. State trajectories of reactants, R, and products, P, are evaluated by the use of a thermodynamic equilibrium algorithm, such as STANJAN (Reynolds 1996), specifying all the coefficients for (4.24).

5. Time profile of the effective mass fraction of products generated in the course of the exothermic stage, $y_E(\Theta)$, is then evaluated by means of (4.24), identifying, by its maximum, the terminal state of the exothermic stage, **t**, $y_{Emax} = y_{Et}$.

6. Profile of the mass fraction of consumed fuel, $x_F(\Theta)$, is, on this basis, determined by means of the correlation function specified by (4.29).

7. With the parameters α_E and χ_E, evaluated by regression of $y_E(\Theta)$, to a life function, profiles of the rate of fuel consumption, $\dot{x}_F(\Theta)$, and its rate of change, $\ddot{x}_F(\Theta)$, akin, respectively, to velocity and acceleration, are determined analytically by (4.43) and (4.44).
8. Temperature and density profiles of reactants, R, and products, P, are established on the basis of (4.21) and (4.22), taking into account their molecular masses specified by the thermodynamic equilibrium algorithm, such as STANJAN (Reynolds 1996).
9. Concentration histories of all the constituents of the exothermic reaction, including prominently the evolution of pollutant species, are evaluated by integration of the set of nonlinear, ordinary differential equations of chemical kinetics: the mass source expressed by (4.46) and the thermal source expressed by (4.51), taking into account the effects of thermal diffusion expressed, in terms of the relaxation time, by (4.52).

5.2 Implementation

Implementation of the diagnostic procedure is illustrated by comparative cases of a diesel and a spark ignition engine. Considered here, in particular, are:

1. Diesel and a spark ignition engines at full load
2. Spark ignition engine at part load
3. Chemical transformations taking place in a spark ignition engine at full and part loads, revealing the evolution of pollutants in the course of combustion in a piston engine.

In all the cases, the mass fraction of reactants was not available. The results correspond, therefore, to $Y_R = 1$ and $\sigma_R = \lambda_{\text{fresh mixture}} \, \sigma_{\text{stoichiometric}}$. The solutions for all of them, moreover, are restricted to the limit of mixed charge, because the ineffective part of consumed fuel can be then ascribed to energy loss by heat transfer to the walls, rather than, as typical of unmixed reactants, to heating the rest of the charge whereby they participate in the exothermic process as the working substance without getting involved in its chemical reaction.

5.2.1 Full Load

The specific powerplants subjected to pressure diagnosis are: a Caterpillar diesel and a spark ignited Renault engine. Their specifications are provided by Table 5.1; the operating conditions of dynamometer tests are specified in Table 5.2, including values of polytropic indexes obtained from pressure measurements according to step 2. The thermodynamic parameters of the components, evaluated by the use of STANJAN (Reynolds 1996), are displayed in Table 5.3 and the coordinates of the diagram of states are presented in Table 5.4 (step 5), whereas the parameters of the dynamic stage of combustion are listed in Table 5.5 (steps 4 and 7).

Table 5.1. Engine data (Gavillet et al. 1993; Oppenheim et al. 1997)

	Diesel (Caterpillar)	Spark Ignition (Renault)
Model	C-12	F7P-700
Bore(mm) x stroke(mm)	130 x 150	82.0 x 83.5
Cylinders	6	4
Piston rod length (mm)	243	144
Compression ratio	16	10

Table 5.2. Operating conditions (Gavillet et al. 1993; Oppenheim et al. 1997)

	Diesel (Caterpillar)	Spark Ignition (Renault)
Speed (rpm)	1200	2000
Torque (Nm)	1944	128
BMEP (kPa)	2037	912
Fuel	$C_{12}H_{26}$	RON 95
Fuel flow (gm/min)	875	32
λ	1.57	1
σ	22.8	15.0
P_i/atm	2.60	1
T_i/K	312	300
n	1399	1323

The measured pressure profiles are depicted in Fig. 5.1. With the time profile of the cylinder volume, $v_s(\Theta)$, specified in an analytic form by kinematics of the crankshaft mechanism, the indicator diagrams, displayed on Fig. 5.2 in linear scales and on Fig. 5.3 in logarithmic scales, are obtained. Determined from the latter, by linear regression, are the polytropic indices of compression, n_c, and expansion, n_e, listed in Table 5.2.

Profiles of the polytropic pressure models, evaluated on their basis, are displayed in Fig. 5.4. The level of the initial point, i, is identified by the base line: $\pi_c(\Theta) = $ const. The final point, f, is specified by two requirements: (1) it corresponds to $\tau_* = $ max and, (2), it is on the expansion polytrope.

Table 5.3. Thermodynamic parameters (Gavillet et al. 1993; Oppenheim et al. 1997)

		p	T	v	u	h	w	M
	States	atm	K	m³/kg		kJ/g		g/mol
Caterpillar								
A	i	124	895	20.5290	0.3755	0.6334	0.2579	28.85
	f	25	586	66.6680	0.1269	0.2958	0.1689	
F	i	124	895	3.4780	0.0922	-0.0485	0.0437	170.33
	f	25	875	16.8687	0.1604	-0.1177	0.0427	
C	i	124	895	19.8155	0.3558	0.06047	0.2489	29.898
	f	25	—	64.6119	0.1148	0.2784	0.1636	
R	i	124	895	19.4635	0.3461	0.5906	0.2445	30.438
	f	25	—	63.5831	0.1089	0.2699	0.1610	
	uv	500	3397	19.4440	0.3458	1.3306	0.9848	28.187
P	hp	124	2885	66.6186	0.2462	0.5906	0.8368	28.369
	pT	124	895	20.6668	2.4696	-2.21	0.2596	28.668
Renault								
A	i	17.18	623	103.12	0.1556	0.3352	0.1795	28.85
	f	44.24	801	51.499	0.2977	0.5285	0.2309	
F	i	17.18	623	26.206	1.2294	-1.1838	0.0456	113,53
	f	44.24	637	10.412	1.1868	-1.1401	0.0467	
R	i	17.18	623	98.328	0.0693	0.24046	0.1712	30.258
	f	44.24	772	47.295	0.2092	0.42124	0.2120	
	uv	141.65	2870	98.328	0.0693	0.9176	0.8483	28.131
P	hp	17.18	2515	423.87	0.4975	0.2405	0.7380	28.330
	pT	17.18	623	104.01	2.7255	-2.5444	0.1811	28.606

Table 5.4. State coordinates (Gavillet et al. 1993; Oppenheim et al. 1997)

K	C_K	U_{k0}	q_K
Caterpillar			
A	2.7892	-0.3439	0
F	70.4601	-3.1705	2.8265
C	2.8201	-0.3461	0.0022
R	2.8363	-0.3473	0.0034
Renault			
P	5.1105	-4.3068	3.9629
A	2.7673	-0.3412	0
F	38.7273	-2.995	2.654
R	3.4281	-0.5176	0.1764
P	5.1373	-4.2886	3.9474

Table 5.5. Bounds of the dynamic stage and life function parameters (Gavillet et al. 1993; Oppenheim et al 1997)

	Diesel (Caterpillar)	Spark Ignition (Renault)
θ_i	360	353
θ_f	410	384
α_π	6.94	9.38
χ_π	1.14	1.29
α_E	5.43	7.53
χ_E	0.87	0.90

With the value of the polytropic exponent, n_e, and fixed initial and final states, the profile of the polytropic model, $\pi_e(\Theta)$, is expressed by regression, in terms of a life function, $x(\tau)$, depicted in Fig. 5.5, whose parameters, α, and χ, are listed in Table 5.5. Deduced thus is an analytical expression for $P(\Theta)$, displayed in Fig. 5.5, demonstrating the remarkable accuracy, with which the life function fits the experimental data. The diagrams of states for the dynamic stage of combustion in the two engines are presented by Fig. 5.6.

The effective mass fraction of generated products, , as well as the effectiveness of the dynamic stage of combustion, y_{Ef}, cited in Table 5.5 are evaluated on this basis, as prescribed in step 5, providing the key to the mass fraction of fuel consumption, determined according to step 6. Profiles of the kinematic parameters, x, and $\tilde{y}_E$, and their corresponding polar diagrams, $\overset{\circ}{\xi}(\dot{\xi})$, are depicted in Figs. 5.7 and 5.8. Finally, profiles of normalized temperatures, $\tilde{T}_K(\Theta) \equiv T_K(\Theta)/T_i$, $(K = R, S, P)$, and normalized specific volumes, $\tilde{v}_K(\Theta) \equiv v_K(\Theta)/v_i$, depicted in Fig. 5.9, following the prescription for step 8.

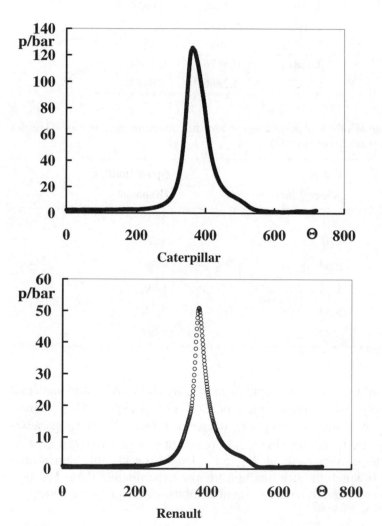

Fig. 5.1. Measured pressure profiles (Gavillet et al. 1993; Oppenheim et al. 1997)

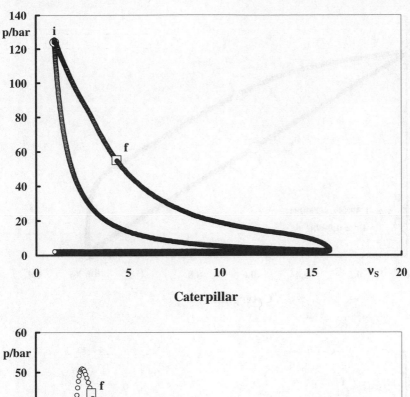

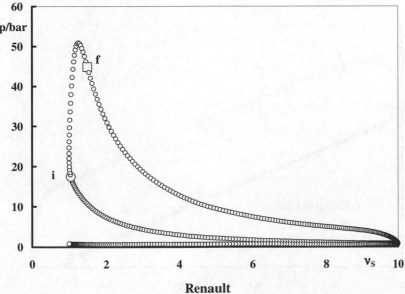

Fig. 5.2. Indicator diagrams in linear scales (Gavillet et al. 1993; Oppenheim et al. 1997)

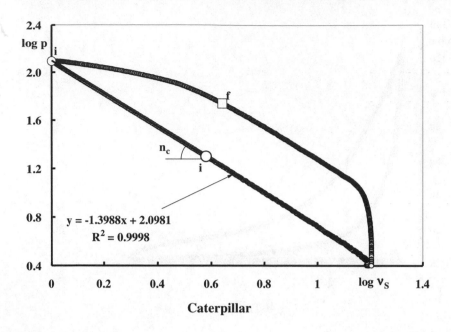

Caterpillar

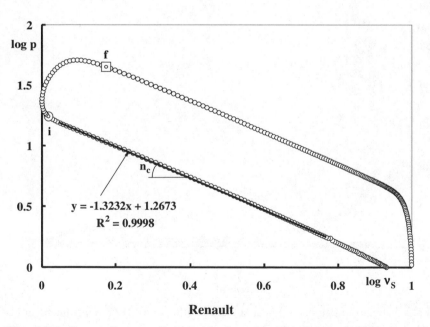

Renault

Fig. 5.3. Indicator diagrams in logarithmic scales (Gavillet et al. 1993; Oppenheim et al. 1997)

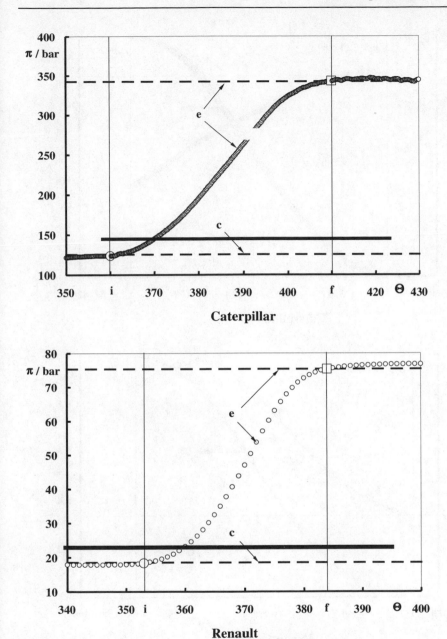

Fig. 5.4. Profiles of polytropic pressure models

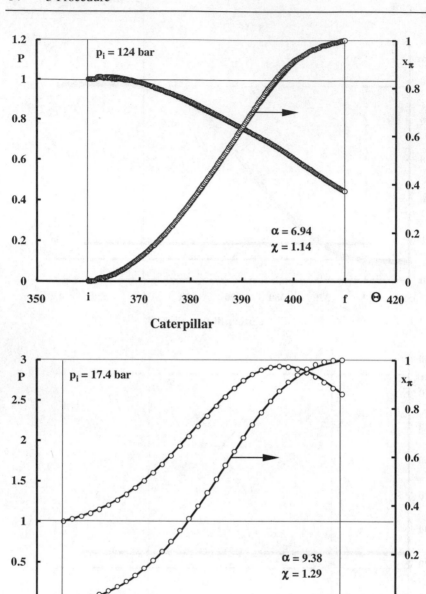

Fig. 5.5. Profiles of measured pressure data, $P(\Theta) \equiv p/p_i$, and the progress parameter, $x_\pi(\Theta)$, in comparison to their analytic expressions, in terms of life functions, of pressure models, displayed by continuous lines

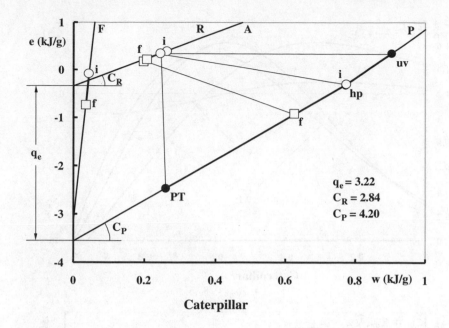

Caterpillar

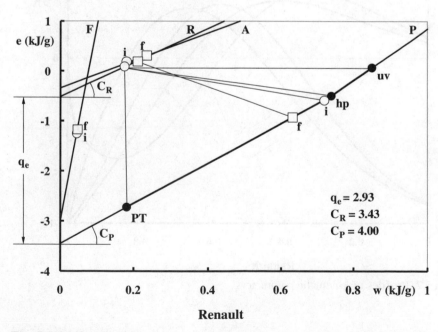

Renault

Fig. 5.6. Diagrams of states

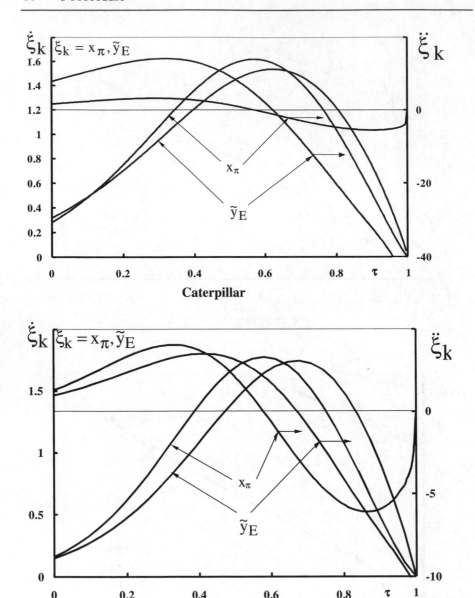

Fig. 5.7. Profiles of kinematic parameters

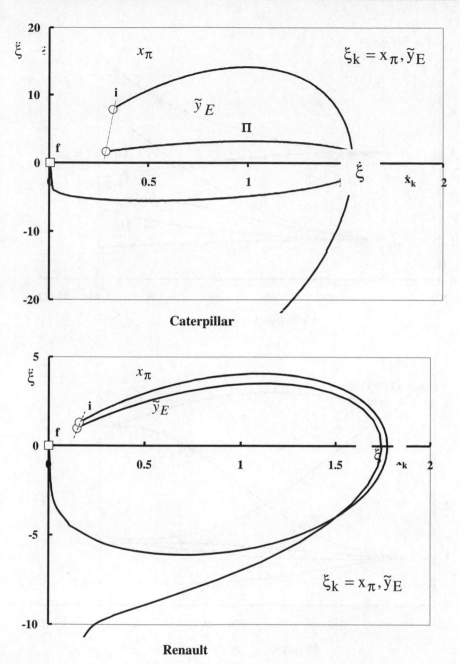

Fig. 5.8. Polar diagrams of kinematic parameters

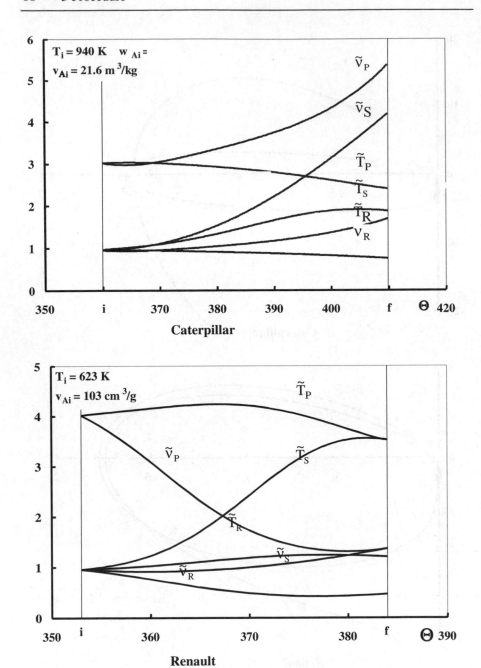

Fig 5.9. Profiles of normalized temperatures and specific volumes

5.2.2 Part Load

For comparison with full load, the performance of the Renault engine was determined also at part load for the most frequently encountered conditions according to the European driving cycle. The data are listed in Table 5.6. Thermodynamic parameters of the components, evaluated by the use of STANJAN (Reynolds 1996), are presented in Table 5.7.

Table 5.6. Operating conditions of the Renault spark ignition engine at part load and its life function parameters (Gavillet et al. 1993)

Speed (rpm)		2000
Torque (Nm)		9.83
BMEP (kPa)		70
Fuel		RON 95
Fuel flow (gm/min)		32
λ		1
σ		15.0
P_i/atm		0.6
T_i/K		300
n		1.365
x_π	α_π	14.2
	χ_π	2.78
y_E	α_E	12.8
	χ_E	2.29

Table 5.7. Thermodynamic parameters of the dynamic stage of combustion in a spark ignition engine operating at part load (Gavillet et al. 1993)

K	States	p	T	v	u	h	w	M
		atm	K	cm³/g		kJ/g		g/mol
A	i	5.56	730	373.50	0.2402	0.4506	0.2104	28.85
	f	5.02	711	402.64	0.2246	0.4294	0.2048	
F	i	5.56	730	94.917	-0.8967	-0.8433	0.0535	113.53
	f	5.02	729	104.89	-0.9021	-0.8487	0.0534	
R	i	5.56	730	356.14	0.1693	0.3699	0.2006	30.26
	f	5.02	714	385.52	0.1536	0.3497	0.1954	
P	uv	23.48	2847	356.14	0.1693	1.0165	0.8472	27.95
	hp	5.56	2534	1326.6	-0.3774	0.3699	0.7473	28.19
	PT	5.56	730	376.70	-2.6298	-2.4176	0.2122	28.61

Indicator diagrams in linear and logarithmic scales are presented by Fig. 5.10. The corresponding profile of the polytropic pressure model, with exponentials n_e and n_c obtained by linear regression, cited in Table 5.6, are depicted in Fig. 5.11, while profiles of the progress parameter $x_\pi(\Theta)$ and of the normalized pressure, $P(\Theta)$, $\equiv p(\Theta)/p_i$, represented by their data (marked by circles) and the analytic expressions derived from the life function, are shown in Fig. 5.12. Kinematic properties of progress parameters are depicted in Fig. 5.13, while the corresponding polar diagrams are displayed in Fig. 5.14. Profiles of normalized temperatures and specific volumes are demonstrated by Fig. 5.15.

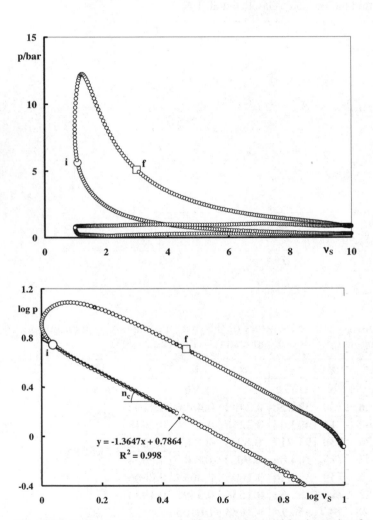

Fig. 5.10. Indicator diagrams in linear and logarithmic scales for Renault engine operated at part load (Gavillet et al. 1993)

Table 5.8. State coordinates of the dynamic stage of combustion in a spark igni-tion engine operating at part load (Gavillet et al. 1993)

K	States		C	u_o
				kJ/g
A	i		2.7857	0.3459
	f			
F	i		54.0000	3.7857
	f			
R	i		3.0192	0.4364
	f			
P	uv		5.4725	4.4670
	hp			
	PT		4.2093	3.5230

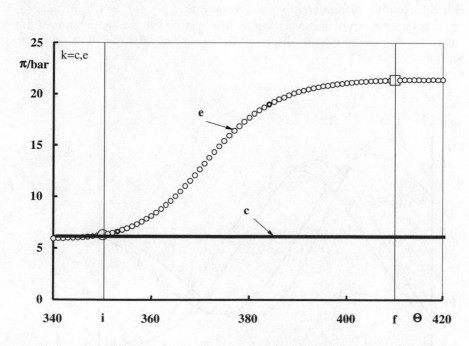

Fig. 5.11. Profiles of polytropic pressure model for Renault engine operated at part load

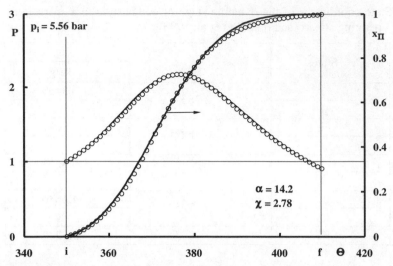

Fig 5.12. Profiles of measured pressure data, $P(\Theta) \equiv p/p_i$, as well as of the progress parameters, $x(\Theta)$, in comparison to their analytic expression, in terms of life functions of pressure models, displayed by continuous lines, for Renault engine operated at part load

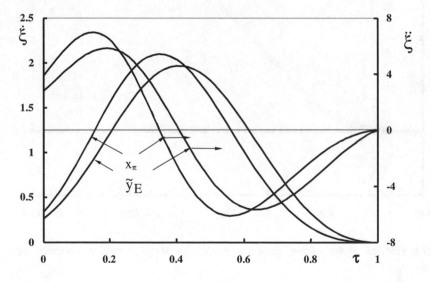

Fig. 5.13. Profiles of kinematic parameters for Renault engine operated at part load

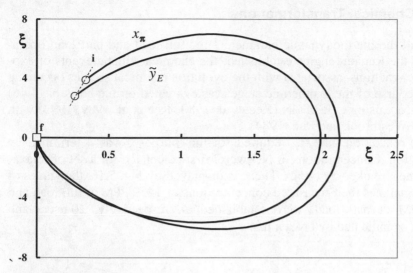

Fig. 5.14. Polar diagrams of kinematic parameters for Renault engine operated at part load

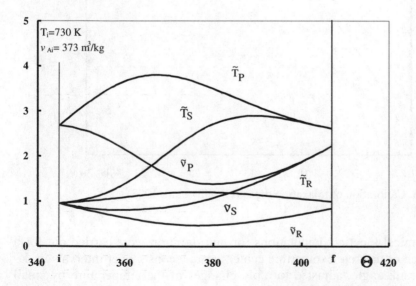

Fig. 5.15. Profiles of normalized temperatures and specific volumes for Renault engine operated at part load)

5.2.3 Chemical Transformations

With all the thermodynamic parameters for full load and part load operation of the Renault engine established, the chemical kinetic events of exothermic reactions, associated with the evolution chemical species occurring
in the course of the exothermic stage was evaluated on the basis of (4.46)
and (4.51), using CHEMKIN (Kee et al. 1980; Kee et al. 1989, 1993) with
data of Westbrook and Pitz (1984).

The relaxation time, τ_T, required for this purpose, was determined by
calibration. Concentrations of NO were, first calculated for a set of relaxation times invoked by (4.52). Then, as displayed by Fig. 5.16, the values of
τ_T that matches their measured concentrations of 1459 PPM at full load and
357 PPM at part load, were established at, respectively, 20 µsec, and
4 µsec, as indicated by broken lines.

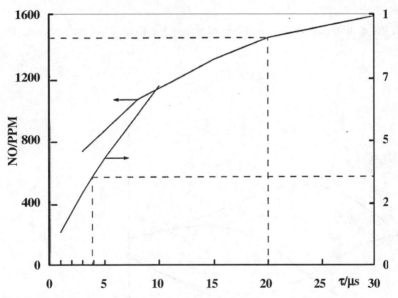

Fig 5.16. Calibration of relaxation times (Gavillet et al. 1993)

Chemical kinetic computations for auto-ignition were carried out for a
sequence of discrete exothermic centers at x_F = const., over intervals in degrees crank angle, adjusted to rapid changes of the temperature by small
steps, and to its slow changes by large steps.

The temperature profiles of representative exothermic centers are presented by Fig. 5.17. The concentration profiles of NO and CO are displayed, respectively, in Figs. 5.18 and 5.19. Their mass average values, of

concentrations, $c_k = \int_i^f y_k \mathrm{d}x_P$, for k = NO, CO, are presented by Fig. 5.20, where, indicated by broken line segments next to the appropriate coordinate axes, are their measured engine-out values - evidently in quite a satisfactory agreement with the results of the calculations.

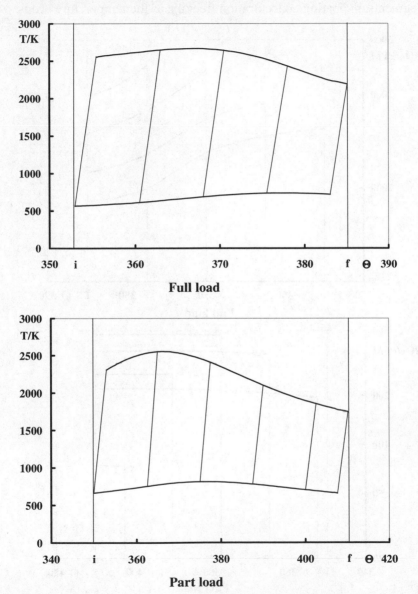

Fig. 5.17. Temperature profiles in representative exothermic centers (Gavillet et al. 1993)

To provide an insight into the thermochemical mechanism of chemical transformations, depicted in Fig. 5.21 are projections onto the T- $\log(c_k)$ plane of integral curves in the multidimensional phase space for $k = NO$ and $k = CO$. Time progresses from the initial state at $-\infty$ in the directions indicated by arrows. Particularly noteworthy in these diagrams are vertical segments, indicating concentration freezing as the temperature drops.

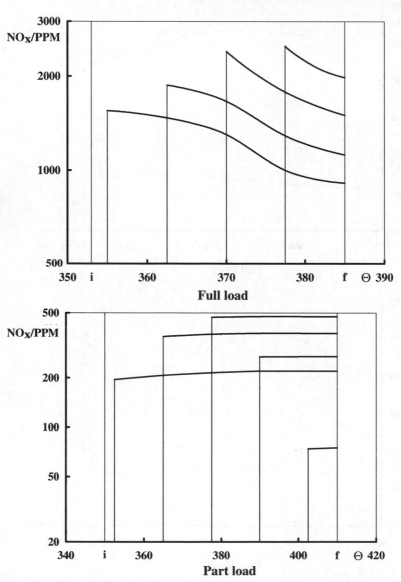

Fig. 5.18. Mole fraction profiles of NO in representative exothermic centers (Gavillet et al. 1993)

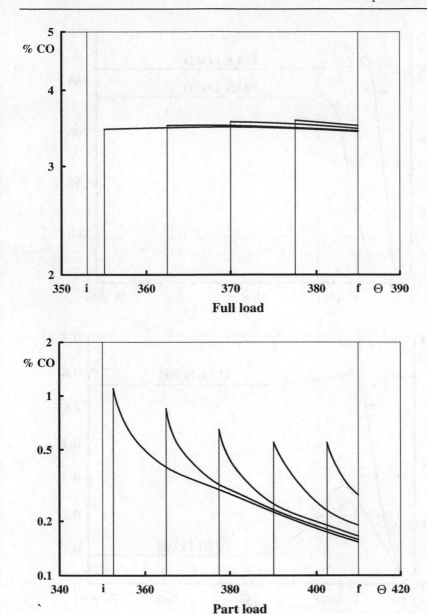

Fig. 5.19. Mole fraction profiles of CO in representative exothermic centers (Gavillet et al. 1993)

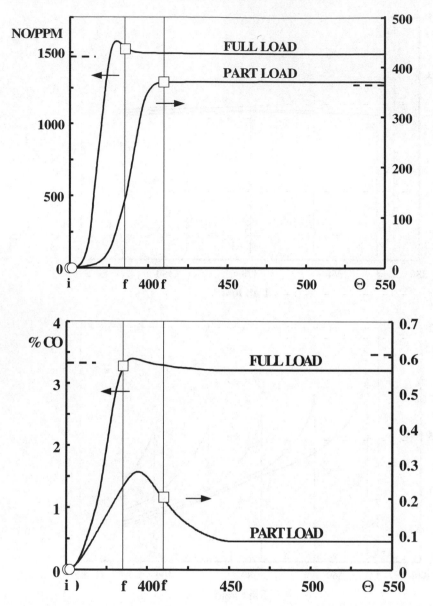

Fig. 5.20. Profiles of mass-averaged mole fraction of pollutants (Gavillet et al. 1993)

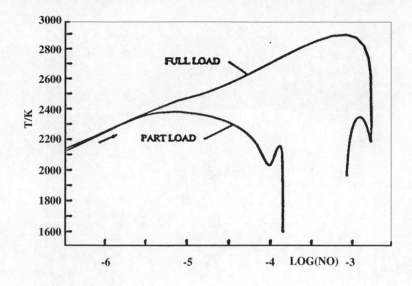

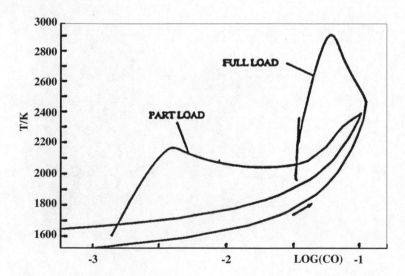

Fig. 5.21. Projection of integral curves pertaining to NO and CO for an exother-mic center (Gavillet et al. 1993)

6 Prognosis

6.1 Background

Pressure diagnostics furnishes a fundamental procedure for design analysis of a combustion system for piston engines. It ushers in thereby a rational method of approach to the prediction of improvements in performance that can be attained by any contemplated modifications of the system prior to their design and testing. In view of the highly developed technology of modern internal combustion engine, such improvements can be derived only from the principles of the Second Law of Thermodynamics, rather than the First Law, in particular reduction in irreversible effects due to energy loss incurred by heat transfer to the walls.

Modifications of the exothermic process of combustion are referred to as internal (or in-cylinder) treatment, in contrast to the external treatment that today is in universal use for reduction of pollutant emissions by employing a chemical processing plant, such as the catalytic converter, in the exhaust pipe. The technology of internal treatment is based on the advantages attained by having the exothermic process executed at a minimum allowable temperature – a condition achievable by reduction of heat transfer to the walls in the course of the dynamic stage, the most critical period in its evolution. This is related directly to the gains accrued by the minimization of unavailable energy (maximization of exergy) – an action that leads to a significant reduction in the formation of chemically generated pollutants, NO_x and CO, as brought out at the end of Chap. 3.

6.2 Project

The project considered here is concerned with evaluation of improvements that can be attained by the Renault 7P engine if, instead of being run in a conventional, throttled, Flame Traversing the Charge (FTC) manner, it is operated in a stratified charge, wide open throttle (WOT), Fireball Mode of Combustion (FMC) (Oppenheim et al 1994).

The operating conditions to be examined are listed in Table 6.1, where Q expresses the net energy loss incurred by heat transfer to the walls, while λ_o and λ_r are the air excess coefficients in the cylinder charge and in the chemically reacting mixture, respectively. Identified thus are six cases of FMC, besides the reference case of FTC, denoted by 0. Cases 1, 2, 3, are concerned with the effects of diminished heat transfer loss, achievable by reducing the contact of the reacting medium with the walls of the cylinder-piston enclosure, while the time interval within which fuel is consumed remains unchanged. Cases 4, 5, 6, take into account, moreover, the consequences of having the lifetime of the dynamic stage is reduced by a factor of two, deemed attainable by increase in combustion rate due to turbulent mixing induced by jet injection and jet ignition.

Table 6.1. Operating conditions (Oppenheim et al. 1994)

	θ_i°	θ_f°	Q	λ_o	λ_r	CASE
FTC	335	410	Q_H	1	1	0
	335	410	Q_H	4.44	1.05	1
	335	410	$Q_H/2$	8.00	1.33	2
	335	410	0	9.09	2.00	3
FMC	335	392,5	$Q_H/2$	5.41	1.67	4
	335	392,5	$Q_H/4$	6.90	2.00	5
	335	392,5	0	9.52	2.50	6

In all the cases, the indicated power is maintained carefully at the same level, as displayed by the indicator diagrams of the FTC and FMC modes of combustion portrayed in Fig. 6.1, corresponding to initial and final states of the dynamic stage, θ_i and θ_f adjusted for maximum IMEP, referred to popularly as MBT (Maximum Brake Torque).

The thermal relaxation time is fixed at $\tau = 5$ μs, as appropriate for part-load operation of the Renault engine according to Fig. 5.18. The computations were based on the detailed kinetic reaction data for oxidation of propane provided by Westbrook and Pitz [94].

For stratified charge of FMC, the operating conditions were established by an iterative procedure. Upon the evaluation of the overall air excess coefficient, λ_o, to provide the required IMEP, the chemical kinetic calculations of combustion in the fireball were performed for a set of postulated air excess coefficients in the reactants, λ_r. Established then was a critical threshold, λ_r^*. For smaller values of λ_r, the chemical induction time remained at a reasonably low level, while, for $\lambda_r > \lambda_r^*$, computation of the oxidation mechanism at the same temperature ceased to proceed. Marked thus was a sharply defined critical limit of extinction, reached as a consequence of dilution by excess air at a fixed temperature, rather than, as is usually the case, by too low temperature at a fixed composition specified by λ_r. Critical values of the air excess coefficient, λ_r, are listed in Table 6.1, and it is for them that all the results of calculations for combustion in fireballs were made.

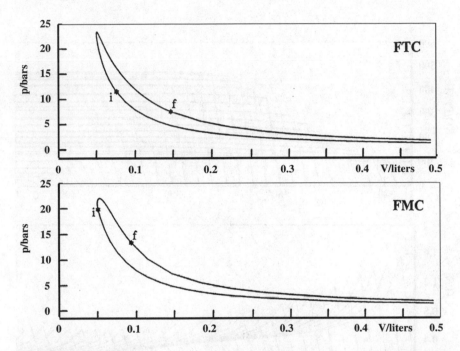

Fig. 6.1. Indicator diagrams for part load operation of the Renault F7P-700 engine at 2000 rpm, operated either by spark-ignited FTC (Flame Traversing the Charge) or in a jet generated FMC (Fireball Mode of Combustion).(Oppenheim et al. 1994)

6.3 Results

Profiles of the temperatures, as well as the concentrations of NO and CO for all the seven cases are presented by Figs. 6.2–6.8 (Oppenheim et al. 1994).

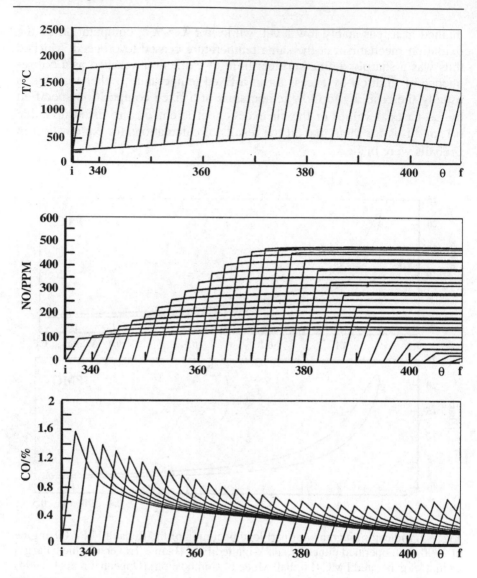

Fig. 6.2. Profiles of the temperature and mole fractions of NO and CO for FTC in case 0 (Oppenheim et al. 1994)

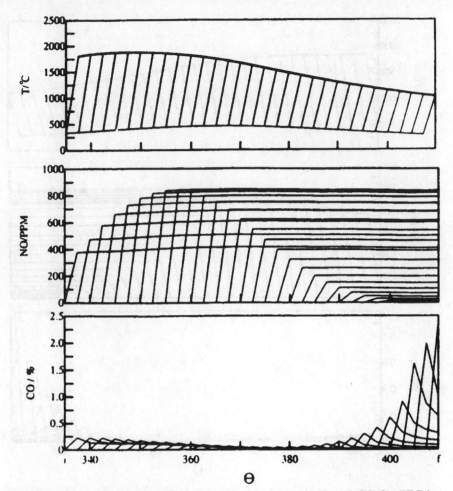

Fig. 6.3. Profiles of the temperature and mole fractions of NO and CO for FTC in case 1 (Oppenheim et al. 1994)

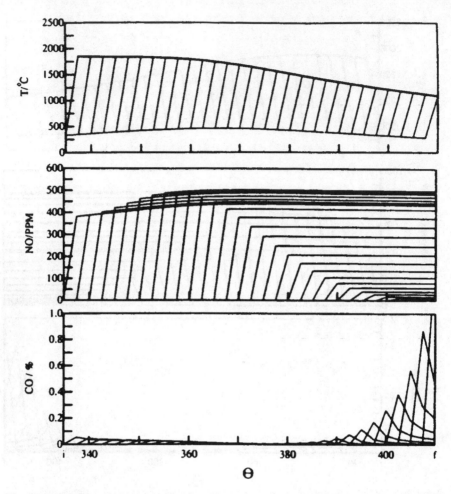

Fig. 6.4. Profiles of the temperature and mole fractions of NO and CO for FTC in case 2 (Oppenheim et al. 1994)

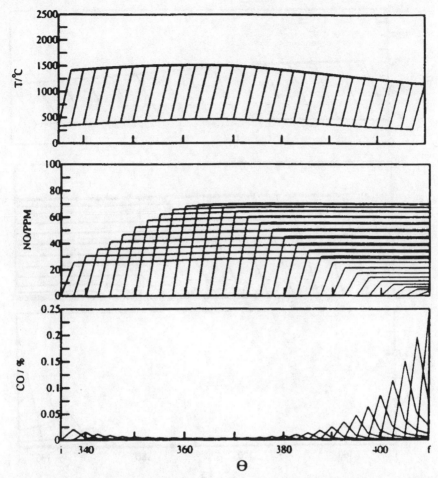

Fig. 6.5. Profiles of the temperature and mole fractions of NO and CO for FTC in case 3 (Oppenheim et al. 1994)

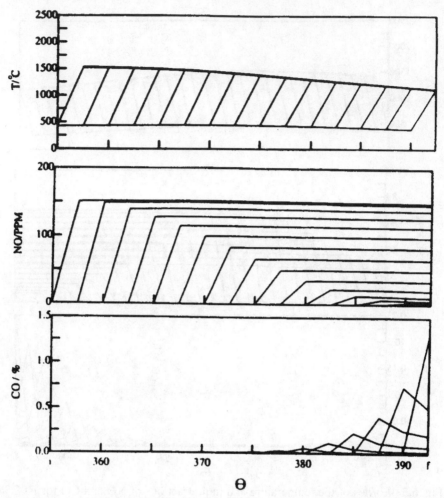

Fig. 6.6. Profiles of the temperature and mole fractions of NO and CO for FTC in case 4 (Oppenheim et al. 1994)

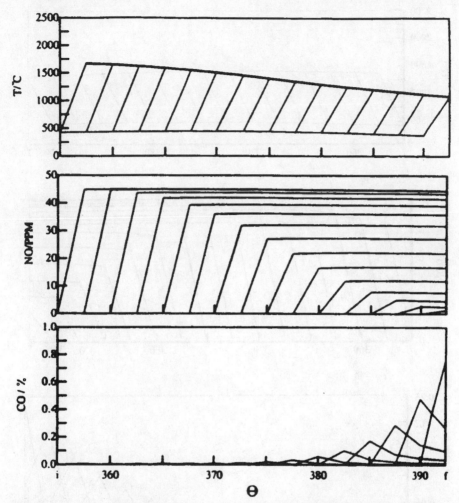

Fig. 6.7. Profiles of the temperature and mole fractions of NO and CO for FTC in case 5 (Oppenheim et al. 1994)

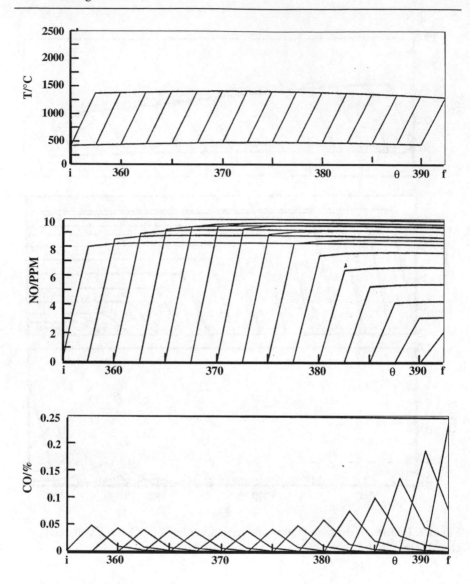

Fig. 6.8. Profiles of the temperature and mole fractions of NO and CO for FMC in case 6 (Oppenheim et al. 1994)

6.4 Conclusions

Principal performance parameters, thus established, are listed in Table 6.2 and displayed in Fig. 6.9.

Table 6.2. Performance parameters (Oppenheim et al. 1994)

CASE	0	1	2	3	4	5	6
ISFC(g/kWh)	388	365	274	170	284	226	165
PPM	368	88	70	8	14	5	1
Mg/g_{fuel}	6.5	5.7	6.3	1.1	1.2	0.6	0.2
ISNO(g/kWh)	2.5	2.1	1.7	0.19	0.35	0.13	0.03
PPM	1600	150	51	10	134	51	19
Mg/g_{fuel}	26.5	9.1	2.5	1.3	11.1	5.4	2.8
ISNO(g/kWh)	5.9	3.3	0.67	0.22	3.2	1.2	0.46

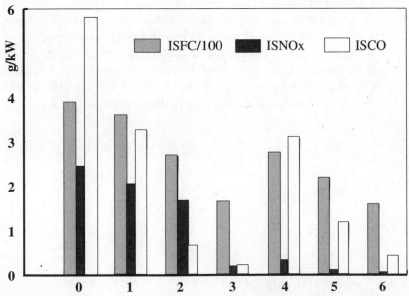

Fig. 6.9. Prognosis of engine performance parameters (Oppenheim et al. 1994)

They present the ISFC[1], ISNO[2], and ISCO[3], obtained for all the cases specified in Table 6.1.

As apparent from them, appreciable dividends are offered by executing the exothermic process of combustion in such a way that the loss of energy caused by heat transfer to the walls is diminished. Principal means to attain these gains are provided by turbulent mixing. The peak temperature can be thereby reduced to a level at which the production of NO is practically annihilated, while, concomitantly, the concentration of CO is cut down by an order of magnitude.

6.5 Résumé

- Presented here is a methodology for evaluating the benefits that can be attained by improved execution of the exothermic process of combustion in a given piston engine.

- A study, carried out for a Renault engine operating at part load, demonstrated that appreciable gains in fuel economy, concomitantly with significant reduction in the formation of chemically generated pollutants, NO and CO, can be achieved by modulating the dynamic stage of combustion so that it is executed away from the walls of the cylinder-piston enclosure, as it can be attained by a jet generated turbulent plume of JIIS[4].

- The advantages accrued by the stratified charge combustion executed in a fireball mode are thus brought out. According to them, the major attribute of internal treatment lies in its ability to have the exothermic process executed at as low temperature as possible by dilution of the charge specified by the overall air excess coefficient, λ_o, as well as of its reacting portion expressed by λ_r.

---

[1] Indicated Specific Fuel Consumption
[2] Indicated Specific NO
[3] Indicated Specific CO
[4] Jet Injection and Ignition System

A Evolution of the Correlation Function

A.1 Introduction

The only way the effective mass fraction of fuel consumed by a combustion system enclosed within impermeable walls, x_E, can differ from its total amount, x_p, upon due account taken of the unburned fuel is expressed in terms of the mass fraction of reactants, Y_R, is by heat transfer. Described here is an experimental and analytical study, carried out for the evaluation of this effect (Oppenheim and Kuhl, 2000a, 2000b).

The experiments involved combustion of lean air/fuel mixtures in a constant volume cylinder fitted with optically transparent sides for schlieren cinematography. The test vessel was equipped with thin film heat transfer probes (surface thermometers) at the walls for time resolved measurements of local heat fluxes, as well as with transducers for recording the concomitant pressure profiles. Heat flux profiles deduced from data recorded by heat transfer probes were found to be, essentially, self-similar and, on this basis, they were integrated yielding profiles of the amount of energy lost by heat transfer to the walls of the enclosure.

As recounted here, mass fraction profiles of fuel consumed in the course of the dynamic stage of combustion are deduced thereupon from the concomitantly measured pressure records by means of the diagnostic procedure described in Chap. 4. A power-law correlation is thus derived between the fuel consumed explicitly to raise pressure, referred to as effective, and its total amount.

As a phase relationship, this correlation is valid irrespectively of the geometry of the enclosure and its variation in a piston engine, especially in the vicinity of the top dead center where the effective consumption of fuel takes place, as well as of the kind of its propagation mechanism, laminar or turbulent flames, or distributed reaction centers like in a HCCI engine. Its applicability is, thus, quite general in providing means for the determination of the ineffective part of consumed fuel, due to energy lost by heat transfer to the walls, from the pressure record, irrespectively of the geometry of the enclosure, its deformation, or the particular manner in which the process of combustion is executed.

Heat transfer in piston engines was, of course, a subject of extensive investigations.[1] Especially prominent in this field are the publications of Woschni[2] – the source of the well-known Woschni correlation. Of particular significance to our studies were the contributions of Greif and his associates.[3] It is on their basis that the reported here self-similarity theory was developed, providing experimental basis for the power law correlation referred to above.

Most of the heat transfer studies were concerned with determination of heat fluxes, as eminently suitable for this subject. For an energy conversion analysis, however, it is not the rate that matters, but the amount, essential for evaluation of the effectiveness with which the exothermic process of combustion is performed.

A.2 Experiments

The experimental investigation was carried out using as a test vessel a cylinder 3.5" in diameter and 2" deep, amounting to 283 cm^3 in volume, closed on both sides by optical glass windows. Its size corresponds to that of a CFR engine cylinder at a compression ratio of 8 : 1, when the piston is at 60 degrees crank angle from the top dead center. For the tests it was filled with a carefully premixed propane-air mixture at equivalence ratio of 0.6, initially at a pressure of 5 bars and a temperature of 65 °C.

Heat flux measurements were made by the use of thin film heat transfer gauges simultaneously with pressure transducer records and high-speed schlieren cinematography taken through the optical plates at both ends of the cylindrical test vessel.

By the use of a variety of ignition systems, the experiments covered a wide range of operating conditions, from an apparently laminar flame to a fully developed turbulent combustion. This was achieved by the use of three distinct modes of executing the exothermic process: (1) spark ignited flame traversing the charge, F, (2) single stream flame jet initiated combustion, S, and (3) triple stream flame jet combustion, T.

The three modes of combustion, F, S and T, are presented, respectively, by schlieren cinematographic records of Figs. 2.11, 2.12 and 2.13.

---

[1] vid. e.g. Annand 1963, Annand and Ma 1970), Krieger and Borman 1966, Le Feuvre et al 1969, Borman and Nishiwaki 1987

[2] vid. e.g. Lange and Woschni 1964, Woschni 1965a, 1965b, 1966/67, 1967, 1970, 1987, Woschni and Anisits 1973, 1974, Woschni and Fieger1979, Csallner and Woschni 1982, Betz and Woschni, 1986, Woschni et al 1986

[3] Heperkan and Greif 1981, Vosen et al 1985, Huang, et al 1987a, 1987b, .Lu et al 1991, Ezekoye et al 1993, Ezekoye and Greif 1993

The four frames displayed in Fig. 2.11 were timed at 5, 10, 20 and 50 msec after the trigger for spark discharge. The frames in all the remaining figures were timed upon a time delay of 21 msec between the trigger for spark ignition in the generator cavity and the record displayed by the first frame, and, thereupon, at 5, 10 and 20 msec later.

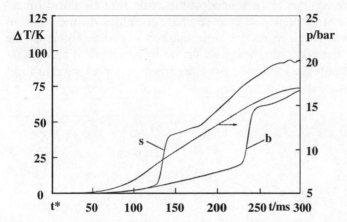

Fig. A.1. Experimental data obtained by thin film thermometric gauges and piezoelectric pressure transducers for the measurement of heat transfer to the walls in mode F of flame traversing the charge

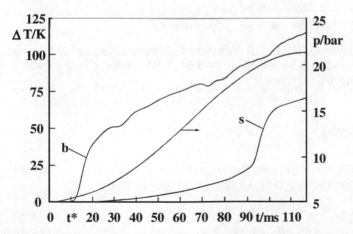

Fig. A.2. Experimental data obtained by thin film thermometric gauges and piezoelectric pressure transducers for the measurement of heat transfer to the walls in mode S of single stream jet ignition

In mode F, the flame front is clearly delineated, acting as the boundary between the reactants and products. Mode S appears as a round turbulent jet plume, while mode T is disk-shaped, in compliance with the geometry of the enclosure. As evident on the photographs, in the first case the products are in contact with the walls right from the outset. In the second, the combustion zone reaches the walls upon a distinct time delay, while in the third this delay is longer.

The temperature profiles, T(t), sensed at the same time by thin film resistance thermometers mounted at two strategic locations on the walls of the cylindrical enclosure, one at the side, marked s, and the other at the back, across from the point of ignition or the jet influx, marked b, together with the concomitantly the measured pressure traces, are presented in Figs. A.1, A.2 and A.3.

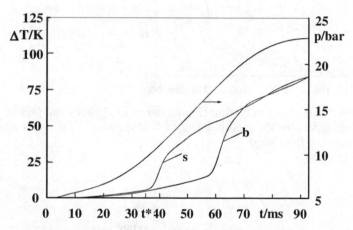

Fig. A.3. Experimental data sensed by thin film thermometric probes and piezo-electric pressure transducers for the measurement of heat transfer to the walls in mode T of triple stream jet ignition

A.3 Heat Transfer

The heat flux was evaluated from the temperature profile, T(t), by the Duhamel superposition integral (vid. Carslaw and Jaeger 1948)

$$\dot{q}''(t) = \beta \int_0^t \frac{dT(t)}{dt_i} \frac{dt_i}{(t - t_i)^{1/2}} \tag{A.1}$$

where $\beta = (k\rho c/\pi)^{1/2}$, while k is the thermal conductivity of the substrate, ρ its density and c the specific heat. For Macor that was employed for this purpose, at the temperature of 300 K, $k = 12.87$ MJ/(s-cm-K), $\rho = 2.52$ g/cm^3 and $c = 0.795$ J/(g-K), so that $\beta = 2.866$ MJ/(s$^{1/-}$-cm^2-K).

The integration was performed using the algorithm of Arpaci (1966). The results are presented by Figs. A.4, A.5 and A.6.

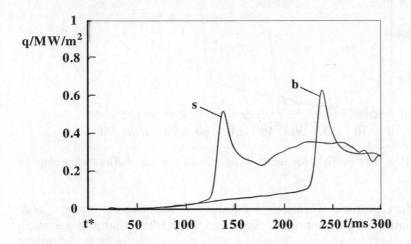

Fig. A.4. Heat flux profiles for mode F deduced from the thermometric data of Fig. A.1

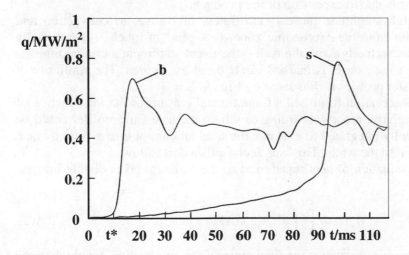

Fig. A.5. Heat flux profiles for mode S deduced from the thermometric data of Fig. A.2

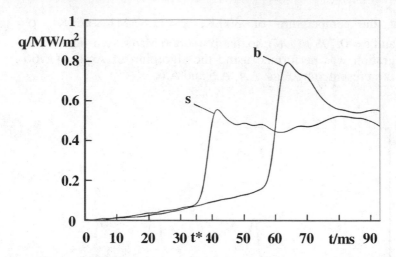

Fig. A.6. Heat flux profiles for mode T deduced from the thermometric data of Fig. A.3

Heat flux profiles at various locations of heat transfer probes, and, hence, starting at different times, displayed a remarkably similar pattern: a ramp followed by a sharp pulse with a wavy decay. The ramp is, according to our estimates, caused predominantly by radiation from the high temperature products, replete of such strong radiators as H_2O and CO_2, while the walls are in contact with the reactants, which are, moreover, heated by compression due to expansion of the products.

The high amplitude pulse is attributed, of course, to convection and conduction from the exothermic zone as it gets "in touch" (operationally rather than factually) with the wall – the event starting at a critical time instant, t^*, whose onset is marked in all the three figures. The similarity of heat transfer profiles is illustrated by Fig. A.7.

This observation furnished a fundamental reason for the introduction of a self-similarity theory according to which the heat flux profiles could be subsequently integrated to evaluate the total amount of energy loss by heat transferred to the walls. This was accomplished as follows.

The evaluation of heat transferred to the walls involves double integration

$$q_w(t) = \int_0^t [\int_0^A \dot{q}''(A, t_i)dA]dt_i \qquad (A.2)$$

where subscript f denotes the final state of the combustion event when the exothermic zone extends over the total wall area of the enclosure.

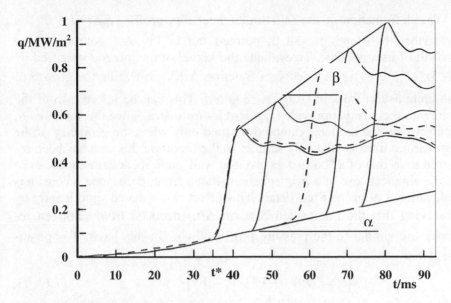

Fig.A.7. Self-similarity features of heat flux profiles

The evolution of heat transfer from a combustion event to the walls of the enclosure consists of two stages.

Up to $t = t^*$, when the exothermic zone gets first "in touch" with the wall at $A = A^*$, the walls are heated only by radiation from the products. Hence, at $t \leq t^*$,

$$q_{wR_o}(t) = A_f \int_0^{t^*} \dot{q}''_{wR}(t_i)dt_i \tag{A.3}$$

At $t \geq t^*$, both the reactants and the products are "in touch" with the walls. For the part of the walls that is not yet "in touch" with the products

$$q_{wR}(t) = \int_*^t [A_f - A(\tau)]\dot{q}''_R(\tau)\,d\tau \tag{A.4}$$

while heat transfer to the part of the walls "in touch" with the products is expressed in terms of a full Duhamel integral

$$q_{wP}(t) = A_f \int_*^t [\int_0^t \dot{q}''_P(t_j;t_i - t_j)\frac{d\Lambda(t_j)}{dt_j}dt_j]d\tau \tag{A.5}$$

where $\Lambda(t_j) \equiv A(t_j)/A_f$, while, with reference to the geometry of the self-similar profiles displayed in Fig. A.3,

$$\dot{q}''_P(t_j;t_i - t_j) = (t_j - t^*)\Lambda(t_j)\,\dot{q}''_P(t^*;t_i - t^*) + [\dot{q}''_R(t_j) - \dot{q}''_R(t^*)] \tag{A.6}$$

The deformation of the self-similar heat flux profile, $\dot{q}_P''(t^*;t_i-t^*)$, is prescribed by slopes α and β, pointed out in Fig. A.7 delineating the growth of its amplitude. To evaluate the kernel of the integral specified by (A.5), $\dot{q}_P''(t_j;t_i-t_j)$, knowledge of function $\Lambda(t_j)$, expressing the growth of the combustion front in (A.6), is required. This can be inferred from the schlieren records that provide fragmentary estimates, since the wall area in contact with the products can be discerned only when the geometry of the exothermic front is relatively simple. In the literature this area has been referred to as that of a "burned-gas wetted wall", and its determination, even in the simplest case of a hemispherical flame front, demanded a considerable amount of geometrical detail. This effort was reduced significantly by observing that the fragmented data on $\Lambda(t_j)$ deduced from schlieren records, are similar to the pressure profile, $P(t_j)$. On this basis, it is postulated that

$$\Lambda(t_j) = [P(t_j) - P^*]/[P_f - P^*] \tag{A.7}$$

where $P^*=P(t^*)$.

The total amount of heat transferred to the walls of the enclosure is given by the sum of the integrals expressed by (A.3), (A.4) and (A.5)

$$q_W(t) = q_{wR_o}(t < t^*) + q_{wR}(t > t^*) + q_{wP}(t > t^*) \tag{A.8}$$

The results of the heat transfer study, expressed in terms of the time profiles of energy loss, q_w/Wh, are presented in Fig. A.8.

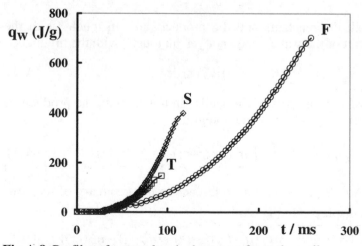

Fig.A.8. Profiles of energy loss by heat transfer to the walls

A.4 Pressure Diagnosis

The measured pressure profiles, depicted in Figs. A.1, A.2, and A.3, are presented on Fig. A.9 in terms of their values normalized with respect to the initial pressure.

Profiles of the life functions, obtained by their regression, are displayed in Fig. A.10. The kinematic properties of the dynamic stages, evaluated on this basis, are depicted in Fig. A.11.

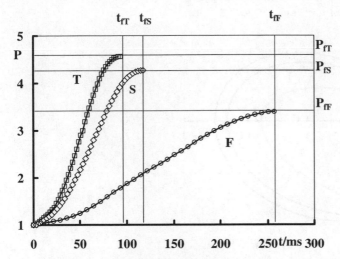

Fig. A.9. Measured pressure profiles in terms of their normalized values

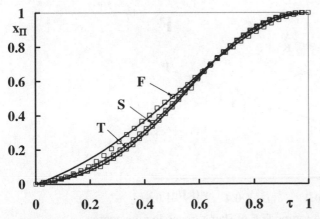

Fig. A.10. Life functions of the measured pressures profiles

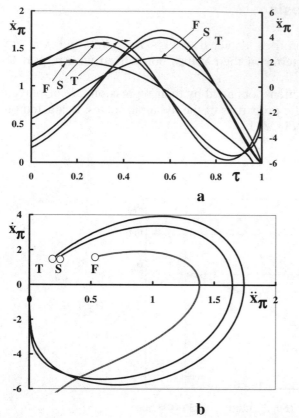

Fig. A.11. Kinematic properties of dynamic stages, expressed in terms of the rates of pressure growth, $\dot{x}_\pi$, and their rates, $\ddot{x}_\pi$. **(a)** time profiles, **(b)** polars

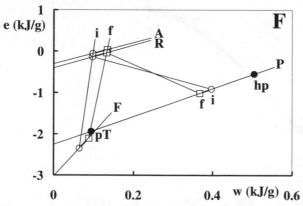

Fig. A.12. State diagram for mode F of flame traversing the charge

The state diagrams of the three modes of combustion, F, S and T, are displayed in Figs. A.12, A.13 and A.14. The effective mass fraction of the generated products, $y_E(t)$, is then evaluated by means of (4.24), whereas their ineffective part, $y_I(t)$, is determined by (4.25), upon the notion that $e_e(t) = q_w(t)$ and, hence, its nominator $Q_I(t) = q_w(t)/w_{Si}$. The results are presented in Figs. A.15, A.16 and A.17, leading, in particular, to the establishment of the mass fraction of reactants, Y_R, the proportionality factor prescribed by (4.2) between the mass fractions of the generated products, $y_L(t)$, (L = P, E, I) and that of the consumed fuel, $x_F(t)$, with its effective and ineffective parts, are presented in Fig. A.18.

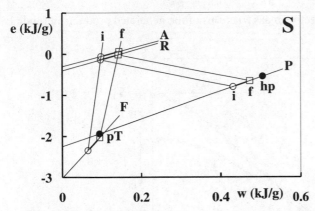

Fig. A.13. State diagram for mode S of single stream jet ignition

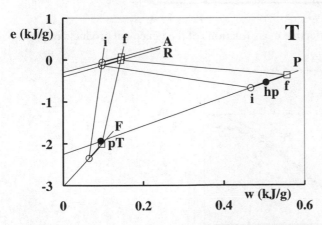

Fig. A.14. State diagram for mode T of triple stream jet ignition

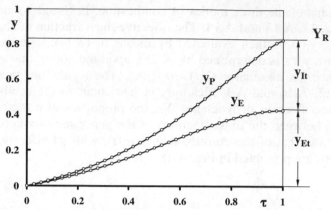

Fig. A.15. Profiles of effective mass fractions of the generated products in mode F

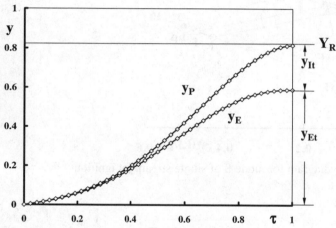

Fig. A.16. Profiles of effective mass fractions of the generated products in mode S

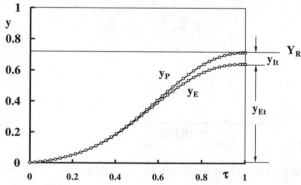

Fig. A.17. Profile of effective mass fractions of generated products in mode T

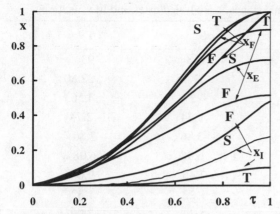

Fig. A.18. Profiles of mass fractions of the consumed fuel

Figure A.19 displays a functional relationship between the mass fraction of consumed fuel, thus obtained, and the progress parameter of the dynamic stage, depicted on Fig. A.10. It illustrates the deviation from the proportionality between them, postulated by Lewis and von Elbe (1987), as well as Rassweiler and Withrow (1938), and promulgated by the well-known "heat release analysis". The data, upon which these figures are based, are listed in Table A.1.

To complete pressure diagnosis carried out upon establishment of Y_R and $x_F(t)$, the kinematic properties of the exothermic stage are displayed on Fig. A.20 and the thermodynamic properties presented by Fig. A.21.

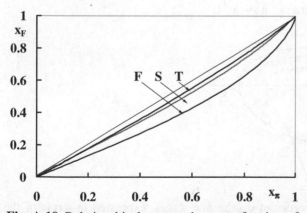

Fig. A.19. Relationship between the mass fraction of consumed fuel, x_F, and the progress parameter of the dynamic stage, x_π.

Table A.1. Parameters of dynamic stages, state trajectories and life functions

Case	F	S	T
T/ms	257	118	93
p_t/bar	16.95	21.35	22.80
P_f	3.39	4.27	4.56
C_P	2.74	2.74	2.74
C_R	3.86	3.86	3.86
Q	6.98	6.98	6.98
α_Π	3.76	7.46	9.00
χ_Π	0.85	1.22	1.32
α_F	2.57	5.73	7.90
χ_F	0.26	0.75	1.06
x_{pf}	0.49	0.71	0.86
η_F	0.87	0.83	0.74

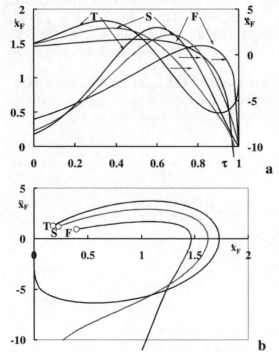

Fig. A.20. Kinematic properties of exothermic stages, expressed in terms of the rates of rate of fuel consumption, $\dot{x}_F$, and its rate of change, $\ddot{x}_F$. **(a)** time profiles, **(b)** polars

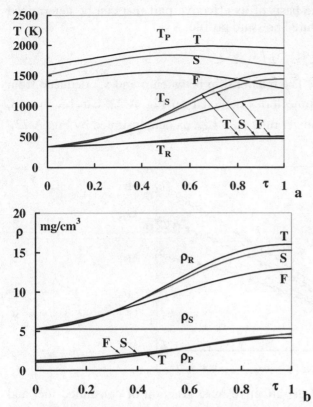

Fig. A.21. Thermodynamic properties of the dynamic stages. **a** Temperature profiles; **b** density profiles

A.5 Correlation

The key to pressure diagnostics is a relationship between the effective mass fraction of consumed fuel, $x_E(t)$, evaluated by virtue of (4.24), and its total amount, $x_F(t)$, presented by Fig. A.18, providing thus the value of the difference between them due to energy loss by heat transfer to the walls.

This is presented by the trajectories on phase plane of $x_E(x_F)$ displayed in Fig. A.22. The correlation between them is expressible in terms of the power function

$$x_E / x_{Et} = y_E / y_{Et} = 1 - (1 - x_F)^\delta \tag{A.9}$$

where $\delta = \sigma/x_{Et}$, while σ is the slope of the trajectory of x_E at $x_F = 0$. The inverse of (A.9) provides the correlation rule for evaluating the mass fraction of consumed fuel on the basis of its effective part that can be determined on the basis of the measured pressure profile

$$x_F = 1 - (1 - x_E / x_{Et})^\kappa \qquad (A.10)$$

where $\kappa = 1/\delta = x_{Et}/\sigma$. The relationship between σ and x_{Et}, deduced from the coordinates of the trajectories depicted on Fig. A.22, can be, in turn, expressed by a simple power rule: $\sigma = x_{Et}^{1/2}$, as demonstrated by Fig. A.23.

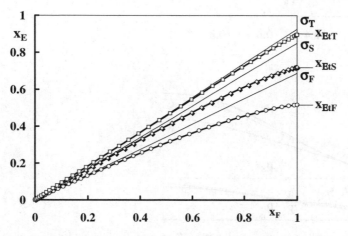

Fig. A.22. Phase plane of the effective mass fraction of consumed fuel and its total amount

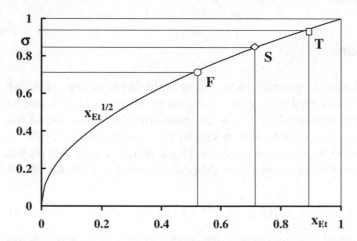

Fig. A.23. Relationship between the power index of phase trajectory of the effective mass fraction of consumed fuel and its maximum value attained at the terminal state of the exothermic stage

Thus, $\kappa = x_{Et}^{1/2}$. Taking into account (4.2), and the first equality in (A.9), the final version of (A.10) is

$$x_F = \frac{y_P}{Y_R} = 1 - (1 - y_E / y_{Et})^{x_{Et}^{1/2}} \qquad (A.11)$$

where the power index, $x_{Et} = y_{Et} / Y_R$. For a given mass fraction of reactants, Y_R, the mass fraction of consumed fuel, x_F, with its kinematic parameters specified by (4.43) and (4.44), is thus obtainable directly from the effective mass fraction of the generated products, y_E, evaluated by means of (4.25).

B Evolution of the Life Function

B.1 Introduction

The essence of combustion is a chemical reaction in the course of which an evolution, associated with transformation of reactants into products, takes place – a metamorphosis affecting usually a part of the system, while the rest changes its thermodynamic state, as a consequence of it, without altering its identity. The system exhibits then all the properties of a dynamic object: its state is displaced from a definite starting point to an end point – a process carried out at a rate, or velocity, whose variation plays the role of acceleration.

Progress of this evolutionary process is recorded, as a rule, by measurements of its symptoms, like concentration of certain species, temperature or pressure, at discrete instances of time, delineating thereby its trajectory determined by interpreting the sampled data points in terms of an integral curve. One is confronted thus with an inverse problem: evaluation of the dynamic properties of the system from the record of the discrete data marking its evolution. Since the dynamic properties are expressed by differentials, finite differences between such data are inappropriate for this purpose. The progress of evolution has to be expressed, therefore, in terms of ordinary differential equations (ODEs). Equations pertaining to evolutionary processes occupy a prominent position in the theory of ODEs. Their formal exposition was provided, among others, by Sell in *Dynamics of evolutionary equations* (1937)and by Hofbauer in *The theory of evolution and dynamical systems* (1956).

Life function and its predecessors presented here are pragmatic examples of solution for such equations, pertaining to specific problems of evolution in biophysics, physical chemistry, and combustion.

B.2 Biophysical Background

The dynamics of evolution is of direct relevance to the mathematical description of life, as manifested prominently in the biophysical literature. Fundamental principles for analytic description of life were formulated by

Lotka in his book on *Elements of Physical Biology*(1924), where he introduced the "Law of Population Growth" in terms of a general equation

$$\frac{dX}{dt} = F(X,t) \tag{B.1}$$

where F represents a prescribed algebraic function. For illustration, he described then a simple, autonomous case of

$$F(X) = aX + bX^2 \tag{B.2}$$

whence, in terms of $\lambda \equiv a/b$, and $X_0 = X$ at $t = 0$

$$X = \frac{\lambda}{(1 + \lambda X_0^{-1})e^{-\alpha t} - 1} \tag{B.3}$$

By adjusting the origin of time at an appropriately selected state, (B.3) can provide an interpretation to a variety of data, as exemplified by Fig. B.1, where, at $t = 0$, $N = \Lambda/2$, so that the population growth

$$N = \frac{\Lambda}{1 + e^{-\alpha t}} \tag{B.4}$$

Another example of the many cases considered by Lotka, is presented here by the growth of a bacterial colony, depicted by Fig. B.2, with its constitutive equation displayed on the diagram.

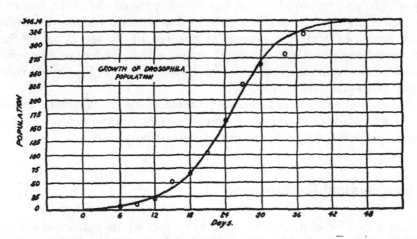

Fig. B.1. A growth curve of population of fruit flies described by Lotka (1924)

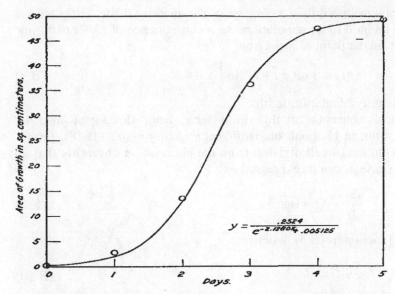

Fig. B 2. Growth curve of a bacterial colony displayed by Lotka (1924)

In a treatise on "Fundamentals of the Theory of the Struggle for Life" ("Les fondements de la théorie de la lutte pour la vie"), Volterra (1937) described the population growth rate of biological species as[1]

$$\frac{dN}{dt} = (a - bN)N \tag{B.5}$$

– an expression that, in terms of a normalized variable $x \equiv (b/a)\,N$, is equivalent to

$$\frac{dx}{dt} = \alpha x(1 - x) \tag{B.6}$$

where $\alpha = a^2/b$. At the two boundaries of $x = 0$ and $x = 1$, $\dot{x} = 0$, while, at $x = \frac{1}{2}$, $\ddot{x} = 0$ - a point of inflection. The time profile of population is expressed then by an S-curve, just like that of (B.4), with this point adopted as the origin of time, described by

$$x = \frac{1}{1 + e^{-\alpha t}} \tag{B.7}$$

so that at $x = 0$ at $t = -\infty$, and $x = 1$ at $t = +\infty$, – an infinite lifespan of god-like quality.

---

[1] vid. Volterra (1937) eq. (1) on p. 4 and its simplified version on p. 5

Volterra introduced then a concept of the quantity if life, $Q(t)$, defined as the time integral of its population. As a consequence of (A.7) with time measured from the point of inflection,

$$Q(t) \equiv \int_0^t x \, dt = t + a^{-1} \ln \frac{1 + e^{-at}}{2} \tag{B.8}$$

– a finite quantity of an infinite life.

Analytical concepts of this kind have been developed also by Rashevski, who, in his book on *Mathematical Biophysics* (1948), formulated the evolution of cell division from the observation observing that, in the simplest case, it can be expressed as[2]

$$\frac{d\varepsilon}{dt} = A\varepsilon - B\varepsilon^2 \tag{B.9}$$

– the same function as (B.5), where

$$\varepsilon = \frac{r_1 - r_2}{\sqrt[3]{r_1 r_2^2}} \tag{B.10}$$

expresses the cell elongation in terms of r_1 – its half-length and r_2 – its half-width.

The process of cell elongation is prescribed then by

$$\varepsilon(t) = \frac{A}{B + A \exp[-A(t - t_0)]} \tag{B.11}$$

--vid. (52) in [9] – an expression equivalent to (A.7).

It is of interest to note that, in their classical paper on the essential mechanism of diffusion, Kolmogorov et al (1937) brought up, as an example of a simplified, one-dimensional problem, the case of the "struggle for life" of a bacteria-like colony. Its evolution was described in terms of the following basic equation[3]

$$\lambda \frac{dv}{dx} = k \frac{d^2 v}{dx^2} + F(v) \tag{B.12}$$

Letting $dv/dx = p$, one gets then

---

[2] vid. Rashevski (1948) eq. (45) on p. 159, with its simplifies version on p.162

[3] vid. 1 Kolmogorov et al (1937) eq. (7) on p.243 in Selected Works (1991)

$$\frac{d^2v}{dx^2} = \frac{dv}{dx}\frac{dp}{dv} = p\frac{dp}{dv} \tag{B.13}$$

and (B.12) becomes

$$\frac{dp}{dv} = \frac{\lambda p - F(v)}{kp} \tag{B.14}$$

- a more general form of the first order ordinary differential equation than (B.1) with (B.2), because here $F = F(X,t)$, rather than just $F(X)$.

B.3 Physico-Chemical Background

It was the advent of the chain reaction theory that gave the impetus to for the generation of mathematical expressions to describe the evolution of physico-chemical processes. The founder of this theory, Nickolai Nickolaevich Semenoff (or Semenov), who got a Noble Prize for its development, provided its detailed description in his monographs on chemical kinetics and reactivity (1934). In the simplest case of a gaseous substance, undergoing an exothermic process in an enclosure of fixed volume, his method of approach is as follows[4].

As a consequence of chain branching, the rate at which a quantity, x, of the substance reacts in time t can be expressed in terms of

$$\frac{dx}{dt} = Bx(P - x) \tag{B.15}$$

– just like (B.6) and (B.9), whereas B is here a measure of the number of chain branching steps per active center, while P is the normalized pressure change associated with this event, upon the understanding, of course, that the reaction takes place in an enclosure of fixed volume.

Thus, in terms of the time interval to reach maximum reaction rate, $\Theta = t - t_m$, where t_m is t at $(dx/dt)_{max}$, just like (B.4), (B.7) and (B.11), it follows that

$$x = \frac{P}{1 + e^{-\phi\Theta}} \tag{B.16}$$

where $\phi \equiv BP$. A plot of this function, (in terms of $\xi = 100x$) and its derivative, pre sented by Semenoff, is reproduced in Fig. B.3.

---

[4] vid. Semenoff 1934 eqs. (49)-(50) pp. 57-68 and Figs. 13 & 14

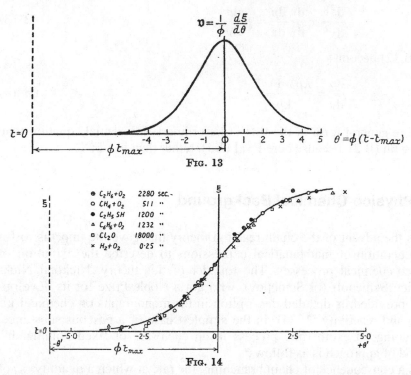

Fig. B.3. Profiles of the amount of substance that has reacted and reaction velocity in comparison to experimental data, provided by Semenoff (1934)

The dynamic features of the auto-catalytic chain reactions were expressed by Kondrat'ev (1964) in the general form of[5]

$$\frac{dx}{dt} = k(x + x_o)^n (1 - x)^m \tag{B.17}$$

whence, in the particular case of $n = m = 1$, upon integration with initial condition of $x = 0$ at $t = 0$,

$$x = x_o \frac{\exp[(1 + x_o)kt] - 1}{1 + x_o \exp[(1 + x_o)kt]} = x_o \frac{1 - \exp(-t')}{x_o + \exp(-t')} \tag{B.18}$$

---

[5] vid. Kondrat'ev (1964). Chapter 1 General Rules for Chemical Reactions; §3. Catalysis by End Products, pp. 38–43, eqs. (3.22), (3.23); (3.25) and Chapter 9 Chain Reactions; §39. Reaction Kinetics Taking into Account Fuel Consumption. Overall Law of the Reaction; p. 624, eqs. (39.35)],

where $t' = (1 + x_0)kt$, while $x = 1$ at $t = \infty$ - an unlikely feature of real life. The solution is thus expressed by an S curve whose point of inflection is located at $t^* = - \ln x_0/[k(1 + x_0)]$. The curve, its slopes, as well as the phase portrait of the solution, are displayed on Fig. B.4.

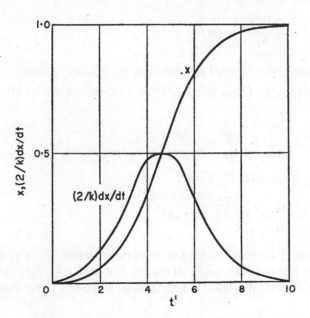

Kinetic curves of an autocatalytic reaction ($x_0 = 0.01$).

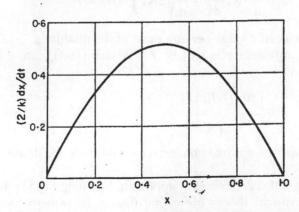

The rate of an autocatalytic reaction as a function of the relative quantity of reacted substance (x).

Fig. B.4. Kinetic rules for chemical reactions according to Kondrat'ev (1964)

B.4 Combustion Background

A similar situation arose in the theory of flame structure, as exposed first by Zel'dovich and Frank-Kamenetskii (1938). According to the Fourier equation of heat conduction in one-dimensional form, they express the thermal flame propagation as

$$K_1 \frac{d^2 T}{dx^2} = -\dot{Q}(T) \tag{B.19}$$

where K_1 denotes the thermal conductivity of reaction products, while $\dot{Q}$ is the volumetric rate of heat release. Then, in terms of $p = dT/dx$, similarly as for (B.13),

$$\frac{d^2 T}{dx^2} = p \frac{dp}{dv} = \frac{1}{2} \frac{dp^2}{dT} \tag{B.20}$$

so that, by quadrature, (B.19) yields

$$\frac{dT}{dx} = \sqrt{\frac{2}{K_1} \int_T^{T_1} \dot{Q}(T) dT} \tag{B.21}$$

Considering T to express the temperature in terms of its progress parameter within the reaction zone, so that at x = 0, T = 0, they derived, by integration of (B.21), the following expression for the mass rate of combustion

$$u = \frac{1}{q} K_1 \frac{dT}{dx} = \frac{1}{q} \sqrt{2K_1 \int_0^{T_f} \dot{Q}(T) dT} \tag{B.22}$$

where q is the calorific value per unit mass of the mixture.

Later, this argument was cast by Zel'dovich (1941) into a more conventional form of[6]

$$u = \frac{[2\eta \int \dot{Q}(T) dT]^{1/2}}{\rho q} \tag{B.23}$$

where   expresses the thermal conductivity, while   is the density of the reactants.

On the basis of this method of approach, Spalding (1957) developed a "temperature explicit" theory of laminar flames. Its principal variable is a generalized progress parameter, known today as the Zeldovich variable,

---

[6] vid. Zeldovich (1941) eqs (15) &(16)

$$\tau \equiv \frac{T - T_i}{T_f - T_i} = Y \tag{B.24}$$

where subscripts i and f denote , respectively, the initial and final states. As a consequence of it, the diffusion and energy equations collapse into one.

Postulating its rate of change across the flame width, ξ, to be described by

$$\frac{d\tau}{d\xi} = \tau(1 - \tau^n) \tag{B.25}$$

– an extended version of (B.5) – he obtained by quadrature the following expression for the flame structure

$$\xi - \xi_0 = \ln \frac{\tau}{(1 - \tau^n)^{1/n}} \tag{B.26}$$

according to which $\tau = 0$ at $\xi = -\infty$, $\tau = 1$ at $\xi = +\infty$, while, at $\xi = \xi_0$, $\tau = 1/2n^{1/n}$ - again an S-curve of infinite life span, as unrealistic as (B.4), (B.7), (B.11) and (B.16).

Over the years, there were a number of similar functional relationships put forth in the combustion literature. For example, in investigating the influence the burning speed exerts upon the working cycle of a diesel engine, Neumann (1934), proposed two functions, $x(\Theta)$, where Θ denotes the time normalized with respect to life time. He first is

$$x = (2 - \Theta)\Theta \tag{B.27}$$

whence

$$\dot{x} = 2(1 - \Theta) \tag{B.28}$$

so that, at $\Theta_i = 0$, $x_i = 0$ and $\dot{x}_i = 2$, while, at $\Theta_f = 1$, $x_f = 1$ and $\dot{x}_f = 0$; whereas $\ddot{x} = -2$.

The second is

$$x = (3 - 2\Theta)\Theta^2 \tag{B.29}$$

whence

$$\dot{x} = 6(1 - \Theta)\Theta \tag{B.30}$$

so that, at $\Theta_i = 0$, $x_i = 0$ and $\dot{x}_i = 0$, while, at $\Theta_f = 1$, $x_f = 1$ and $\dot{x}_f = 0$; whereas $\ddot{x} = 6 - 7\Theta$, whence at the point of inflection, $\ddot{x} = 0$, $\Theta^* = 6/7$.

Later, in a publication on a "Precise Method for the Calculation and Interpretation of Engine Indicator Diagrams," Gončar (1954) introduced an empirical formula

$$x = 1 - (1 + \Theta)e^{-\Theta} \tag{B.31}$$

where $\Theta \equiv t/t_m$, is the ratio of elapsed time to the time of the maximum rate of combustion (maximum burning speed at the point of inflection).

According to (A.31), the rate of combustion is

$$\dot{x} = \frac{\Theta}{t_m} e^{-\Theta} \tag{B.32}$$

while its rate of change

$$\ddot{x} = \frac{1}{t_m^2} (1 - \Theta)e^{-\Theta} \tag{B.33}$$

whence the maximum burning speed, i.e. the rate of combustion at the point of inflection,

$$\dot{x}^* = 1 / e t_m \tag{B.34}$$

so that, when $t_m \to 0$, $\dot{x}^* \to \infty$, while, when $t_m \to \infty$, $\dot{x}^* \to 0$.

B.5 Vibe Function

Following the elucidation provided by Erofeev (1946) Vibe (1956, 1970), expressed Semenov's method of approach on the basis of the postulate that the rate at which reacting molecules, N, decay, due to consumption by chemical reaction, is directly proportional to the rate at which the effective reaction centers, N_e, are engendered, i.e.

$$\frac{dN}{dt} = -n \frac{dN_e}{dt} \tag{B.35}$$

while the latter is expressed in terms of a relative number density function, ρ, as

$$\frac{dN_e}{dt} = \rho N \tag{B.36}$$

whence, with t expressed by the progress parameter of time and $N = N_0$ at $t = 0$,

$$N = N_0 \exp\left(-\int_0^t n\rho dt\right) \tag{B.37}$$

Now, if $\rho = kt^m$, while n = const, the fraction of molecules consumed by chemical reaction[7]

---

[7] vid. Vibe (1956) eq. (7), or Vibe (1970) eq. (44)

$$x \equiv \frac{N_0 - N}{N_0} = 1 - \exp(-\frac{nk}{m+1}t^{m+1}) \tag{B.38}$$

The function expressed by (B.38) is specified by two positive parameters, $\alpha = nk/(m + 1)$ and $\beta = m + 1$. With their use (B.38) is normalized, so that at $t = 0$, $x_i = 0$, while, at $t = 1$, $x_f = 1 - e^{-\alpha}$, yielding

$$x = \frac{1 - \exp(-\alpha t^{\beta})}{1 - \exp(-\alpha)} \tag{B.39}$$

– an expression referred to in the English literature on internal combustion engines[8] as the Wiebe function, rather than Vibe – the proper name of its founder, I.I. Vibe (И.И. ВИбе), Professor at the Ural Polytechnic Institute in Sverdlosk (now Ekaterinburg).

It was introduced by him in the nineteen fifties (Vibe 1956) and later, upon many publications described its origin, in his book under the title "Novel Views on the Engine Working Cycle: Rate of Combustion and Working Cycle in Engines" (Vibe 1970). Early publications of Vibe came to the attention of Professor Jante at the Dresden Technical University, who, upon their translation into German by his associate, Frick, wrote an enthusiastic paper entitled "The Wiebe Combustion Law" ("Das Wiebe-Brenngesetz") (Jante 1960). The name, as well as the initials (J.J. rather than I.I.), of Vibe were then misspelled – a misnomer that became thereby inadvertently introduced into English literature by Heywood et al. (1979.

According to (B.39),

$$\dot{x} = \frac{\alpha \beta t^{\beta-1} \exp(-\alpha t^{\beta})}{1 - \exp(-\alpha)} \tag{B.40}$$

whence, for $\beta > 1$, it follows that, at the initial state of $t = 0$, $\dot{x}_i = 0$, while, at the final state of $t = 1$, $\dot{x}_f = \alpha \beta e^{-\alpha} /(1 - e^{-\alpha}) > 0$.

Its rate of change is

$$\ddot{x} = \frac{\alpha \beta t^{\beta-2}(\beta - 1 - \alpha \beta t^{\beta}) \exp(-tt^{\beta})}{1 - \exp(\alpha)} \tag{B.41}$$

so that at $t = 0$, $\ddot{x}_i = 0$, while, at $t = 1$, $\ddot{x}_f = \alpha \beta(\beta - 1 - \alpha \beta)e^{-\alpha}$, whereas $\ddot{x} = 0$ is at

$$t^* = (\frac{1 - \beta^{-1}}{\alpha})^{\beta^{-1}} \tag{B.42}$$

---

[8] vid. e.g. Heywood (1988), as well as Horlock and Winterbone (1986), in contrast to the German book of Pischinger et al(1989-2002) where it is spelled correctly.

Examples of this function, with profiles of its slopes, are displayed in Fig. B.5 for $\alpha = 3$, and $\beta = 1.5$, 3 and 6. As brought out by it, for $\beta > 1$ the Vibe function describes an S-curve, starting at $t = 0$ with an exponential growth at, initially, zero slope, traversing through a point of inflection into an exponentially decaying stage, until it reaches $t = 1$ at a finite, positive slope. For $\beta = 1$, it is reduced to an expression for a straightforward exponential growth, whereas, for $\beta \leq 1$ and $t^* \leq 0$, so that, within the interval 0-1, it depicts only a decaying growth.

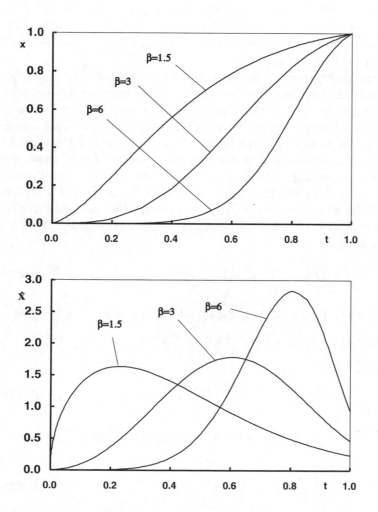

Fig. B5. Time-profiles of the mass fraction of fuel consumed by combustion and its rates for $\alpha = 3$, according to the Vibe function

B.7 Life Function

The Vibe function is a reverse of the life function introduced here in Chap. 4 – it is based on dynamic, rather than physico-chemical, considerations as an integral curve of a differential equation expressing the rate of evolution, or growth.

In contrast to the Vibe function, the life function starts at a finite slope, portraying the dramatic effects of birth, and ends at zero slopes – a state of equilibrium between the positive tendency to grow and the negative influence of decay that is characteristic of natural death taking place upon inflection (menopause). It is presented here, for comparison with the Vibe function, in terms of the same parameters. α and $\beta \equiv \chi + 1$, as those used in (B.39), by the following expression

$$x = \frac{\exp[-\alpha(1-t)^{\beta}] - \exp(-\alpha)}{1 - \exp(-\alpha)} \tag{B.43}$$

Its derivative,

$$\dot{x} = \frac{\alpha\beta(1-t)^{\beta-1} \exp[-\alpha(1-t)^{\beta}]}{1 - \exp(-\alpha)} \tag{B.44}$$

whence, as required[9], at t = 0: x = 0, while $\dot{x} = \alpha\beta/(\exp(\alpha)-1) > 0$, whereas at t = 1: x = 1, while $\dot{x} = 0$.

To display the salient features of the life function in comparison to the Vibe function, plots of (B.43) and (B.44) are shown on Fig. B.6, for the same values of parameters   and   as those of Fig. B.5.

From the differential of (B.44)

$$\ddot{x} = \frac{1 - \beta + \alpha\beta(1-t)^{\beta}}{1-t} \dot{x} \tag{B.45}$$

it follows that the point of inflection, corresponding to $\ddot{x} = 0$, is located at

$$t^{*} = 1 - (\frac{\beta-1}{\alpha\beta})^{1/\beta} \tag{B.46}$$

and

$$x^{*} = \frac{\exp(\beta^{-1}-1) - \exp(-\alpha)}{1 - \exp(-\alpha)} \tag{B.47}$$

---

[9] vid. Sect. 4.9

while its slope there is prescribed by

$$\dot{x}^{*} = \frac{\beta - 1 - \exp(\beta^{-1} - 1)}{1 - \exp(-\alpha)} \qquad \text{(B.48)}$$

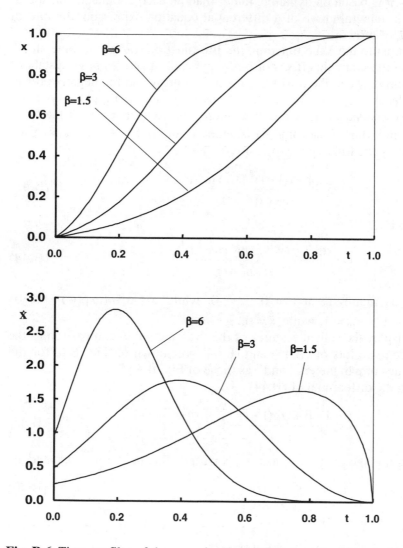

Fig. B.6. Time-profiles of the mass fraction of fuel consumed by combustion and with its rates for $\alpha = 3$, featuring trajectories of constant parameter, β (Oppenheim and Kuhl 1998)

Summary

Part 1

- The technology of combustion in piston engines is today at the threshold of a quantum leap, comparable to that which took place in electronics by its transition from vacuum tubes to transistors – a break-through due to recognition of the exact nature and function of electrons in transmitting electricity. In combustion, the equivalent role to electrons is played by active radicals – the chain carriers in a chemical reaction mechanism. It is the control of their action in the course of an exothermic (energy generating) process that provides the key for a similar evolution to take place in the engine combustion systems.
- An engine cylinder is the site where the working substance is at a high temperature. In principle, therefore, it ought to provide optimum conditions for its chemical reaction. However, as a consequence of its excessively high level, the reaction has been deemed too fast for external control.
- Hence, the industry had to resort to internally administered chemical additives, such as hydrogenated and oxygenated hydrocarbons, exemplified prominently, on one side, by high octane and cetane number fuels, and, on the other, by the notorious MTBE. Operating the exothermic process of combustion at lower temperatures by turbulent jets actuated by impressively high speed modern electronic control systems offers means to accomplish what hitherto has been considered out of question.
- Ushered in, thereby, should be the technology of internal treatment, where the task of minimizing the generation of pollutants is in synergy with that of maximizing fuel economy. This is in contrast to the technology of external treatment that is in universal practice today, where the two tasks are in mutual conflict – a circumstance demanding a trade-off.
- To generate pressure for production of motive work, the fuel is consumed within the exothermic stage – a process whose duration (lifetime) is of an order of one tenth of a revolution. In the cylinder of an

engine operating at cruising speed of an automobile, it takes just a couple of millisecond.

- To acquire insight into a millisecond event, microsecond resolution is required – an appropriate order of magnitude for effective action of an active radical. Progress in the technology of combustion in piston engines must encompass ushering in Microtechnology into its modus operandi.

- Advances in the technology of combustion should be conducted methodically, being executed step-by step, taking full advantage of the best that modern computer technology can offer.

- To start with, proper background should be laid down by assessing the effectiveness with which fuel is utilized in a contemporary engine. Since pressure is the essential product of the combustion system, this is accomplished principally by pressure diagnostics - the solution of an inverse problem in deduction of information on how an action has been executed from its recorded outcome – a task for which, normally, an abundance of data is available from transducer measurements.

- Thereupon, the development of combustion systems should be concerned with proper management and control of their essential components: fuel and air, as they are mixed to form the reactants, and become involved in an exothermic chemical reaction generating the products.

- Progress should be fostered, therefore, by ushering in MEMS (Micro-Electro-Mechanical System) technology. Injection of fuel, its mixing with air and ignition would be thus executed by micro-electronically-operated turbulent jets.

- The exothermic process of combustion would be carried out then under control of an electronic system, the essential feature of MECC (Micro-Electronically Controlled Combustion).

- Prospective benefits of MECC should be of substantial significance to the world economy. By proper control of the exothermic process of combustion, its action as a generator of pollutants can be vastly reduced, while its effectiveness as energy converter can be greatly improved. Mobile sources of pollution can be thereby virtually eliminated, while the mileage out of a tank of fuel at least doubled.

Part 2

- The method of pressure diagnostics for assessing the evolution and effectiveness of the exothermic process of combustion in a piston engine is presented. In view of its dynamic character, the product of pressure and specific volume, $w \equiv pv = p/\rho$, expressing the mechanical properties of force and displacement, is adopted as the principal thermodynamic reference parameter. The temperature, employed conventionally for this purpose, is thus relegated to secondary role of a dependent variable, which can be evaluated, whenever required, from the appropriate equation of state.

- The fundamental parameter of state complementing w, is the internal energy per unit mass, e. A Cartesian plane, formed by them as its coordinates, is the platform, at a reference pressure level, of a three-dimensional phase space for a complete set of data on the thermodynamic states of all the components participating in the combustion event. The cylinder charge and the reactants are treated then as a mixture at a fixed chemical composition that is compressed and mixed, whereas the products of reaction are at their equilibrium states.

- For a closed system, the balances of mass, volume and energy provide a functional relationship between the evolution of the mass fraction of the generated products and the measured pressure profile. As the latter is obtained from measurement, accomplished thereby is the task of pressure diagnostics – the solution for inverse problem of deducing information on the behavior of a system from its measured symptoms.

- Specific interpretation is thus obtained of the evolution of the exothermic stage of combustion, yielding data on the thermodynamic processes taking place in its course, on the profiles of mass fractions of the system components, and, in particular, on the effectiveness with which fuel is consumed to generate pressure – its sole raison d'être.

- The exothermic process of combustion is considered for this purpose as a dynamic system to comply with requirement for design of a micro-electronic control apparatus. In this respect, the mass fraction of consumed fuel is akin to distance, its rate of consumption akin to velocity and the change of this rate akin to acceleration.

- Interpretation of molecular transformations taking place in the course of the exothermic process of combustion is, thereupon, determined by chemical kinetic analysis of the transformation of reactants onto products between their initial and final states established by the thermodynamic analysis. Obtained then, in particular, are data on the formation of all the pollutants.

- A fundamental background for the development of micro-electronic control system for execution of the exothermic process of combustion in piston engines is thus provided.

- To culminate Part 2, here is another rendition of Cal Bear expressing the joy and satisfaction he got out of his caldron by taking advantage of the best that culinary art and science can offer.

References

Annand WJD (1963) Heat transfer in the cylinders of reciprocating internal combustion engines. Proceedings of the Institution of Mechanical Engineers 177: 973–996

Annand WJD, Ma T H (1970) Instantaneous heat transfer rates to the cylinder wall surface of a small compression-ignition engine. Proceedings of the Institution of Mechanical Engineers 185: 976–987

Arpaci VS (1966) Conduction Heat Transfer, esp. pp. 307–308, Addison-Wesley

Ashley S (2001) A Low Pollution Engine Solution, Scientific American: 91–95

Benson RA (1962) The thermodynamics and gas dynamics of internal-combustion engines (ed. J.H. Horlock and D.E. Winterbone) Clarendon Press, Oxford

Benson SW (1960 The foundations of chemical kinetics. McGraw-Hill Book Co., New York, xvii +703 pp

Benson SW (1976) Thermochemical kinetics. John Wiley & Sons, New York, xi + 320 pp

Betz A, Woschni G (1986) Umsetzungsgrad und Brennverlauf aufgeladener Dieselmotoren im instationären Betrieb, MTZ 47: 263–267

Blair, GP (1996) Design and simulation of two-stroke engines, SAE, Warrendale

Blair, GP. (1999) *Design and simulation of four-stroke engines*, SAE, Warrendale

Boddington T, Gray P, Harvey DL (1971) Thermal theory of spontaneous ignition: criticality in bodies of arbitrary shape. Phil. Trans. Roy. Soc. Lond. A270: 467–506

Borman G, Nishiwaki K (1987) Internal-combustion engine heat transfer. Prog. in Energy and Combustion Science 13: 1–16

Boyd TA (1950) Pathfinding in Fuels and Engines. SAE Quarterly Transactions.4: 182–195

Bui TD, Oppenheim AK, Pratt DT (1984) Recent advances in methods for numerical solution of O.D.E. initial value problems. J. Computational and Applied Mathematics 11: 283–296

Caratheodory C (1909) Untersuchungen über die Grundlagen der Thermodynamik. Math.Ann. 67: 355–386

Carslaw HS, Jaeger JC (1948) Conduction of Heat in Solids, esp. p.43, Oxford Press, Oxford

Cetegen B, Teichman KY, Weinberg FJ, Oppenheim AK (1980) Performance of a Plasma Jet Igniter. SAE Paper 800042, 14 pp, SAE Transactions 89: 246–259

Csallner, P. & Woschni, G. 1982 Zur Vorausberechnung des Brennverlaufes von Ottomotoren bei geänderten Betriebsbedingungen MTZ 43, 195–200

Curran HJ, Gaffuri P, Pitz WJ, Westbrook CK (1998) A Comprehensive Modeling Study of n-Heptane Oxidation" Combustion and Flame 114: 149–177

Czerwinski J, Spektor R (2000) Application of the Oppenheim Correlation (OPC) for Evaluation of Heat Losses from Combustion in IC-Engine" SAE Paper 2000-01-0202, SP. 1492, Advances in Combustion 2000: 123–134

Dale JD, Oppenheim AK (1981) Enhanced Ignition for I.C. Engines with Premixed Gases," SAE 810146, SAE Transactions 90: 606–621

Dale JD, Oppenheim AK (1982) A Rationale for Advances in the Technology of I.C. Engines. SAE 820047, 15 pp

Edwards CF, Oppenheim AK, Dale JD (1983) A Comparative Study of Plasma Ignition Systems. SAE 830479, 12 pp

Edwards CF, Stewart HE, Oppenheim AK (1985) A Photographic Study of Plasma Ignition Systems. SAE 850077, 10 pp

Erofeev BV (1946) Generalized Equation of Chemical Kinetics and its Application to Reactions involving solid phase components, Doklady AN USSR 51, 6

Ezekoye OA, Greif R, Sawyer RF (1992) Increased surface t effects on wall heat transfer during unsteady flame quenching. Twenty-Fourth International Symposium on Combustion, The Combustion Institute, 1465–1472

Ezekoye, O.A., Greif, R. 1993 A comparison of one and two dimensional flame quenching: heat transfer results ASME HTD, vol. 250 Heat Transfer in Fires and Combustion Systems, 10 pp

Gardiner WC (1972) Rates and Mechanisms of Chemical Reactions. WA Benjamin Inc. Menlo Park, CA, x + 284 pp

Gavillet GG, Maxson JA, Oppenheim AK (1993) Thermodynamic and Thermochemical Aspects of Combustion in Premixed Charge Engines Revisited SAE 930432, 20 pp

Gear CW (1971) Numerical initial value problems for ordinary differential equations, Prentice-Hall, New York, xvii + 253 pp

Gibbs JW (1875–1878) On the equilibrium of heterogeneous substances. Transactions of the Connecticut Academy, III (1875–1876) pp. 108–248; (1877–1878) pp 343–524 (1931) The Collected Works of J.W. Gibbs, Article III, Longmans, Green and Company, New York, 2: 55-353, esp. pp.85–89 and 96–100

Gončar BM (1954) Precision Method for the Calculation and Presentation of Engine Indicator Diagrams (Ein Präzisiertes Verfahren zur Berechnung und Darstellung eines Motorindikatordiagrammes. Sammelbände Untersuchung der Arbeitsprozesse in Dieselmotoren. CNIDI 25, Mašciz

Gordon S, McBride BJ (1994) Computer Program for Calculation of Complex Chemical Equilibrium Compositions and Applications. I. Analysis, NASA Reference Publication 1311, vi + 55 pp

Gray BF, Yang CH (1965) On the unification of the thermal and chain theories of explosion limits. J. Phys. Chem. 69: 2747–2750

Gray P, Scott SK (1990) Chemical oscillations and instabilities (Non-linear chemical kinetics). Clarendon Press, Oxford, xvi + 453 pp

Griffiths JF (1990) Thermokinetic interactions in simple gaseous reactions. Ann. Rev. Phys. Chem. 36: 77–104

Gussak LA (1976) High Chemical Activity of Incomplete Combustion Products and a Method of Pre-chamber Torch Ignition for Avalanche Activation of Combustion in Internal Combustion Engines. SAE Transactions 84: 2421–2445

Gussak LA (1983) The Role of Chemical Activity and Turbulence Intensity in Pre-chamber-Torch Organization of Combustion of a Stationary Flow of a fuel-Air Mixture. SAE 830592

Gussak LA, Karpov VP, Tikhonov YuY (1979) The Application of the Lag-Process in Pre-chamber Engines. SAE 790692

Gussak LA, Turkish MC (1977) LAG-Process of Combustion and its application in automobile gasoline engines. Stratified Charge Engines, The Institution of Mechanical Engineers, London, pp 137–145

Hensinger DM, Maxson JA., Hom K., Oppenheim AK (1992) Jet Plume Injection and Combustion. SAE 920414, SAE Transactions 10 pp

Heperkan H, Greif R (1981) Heat transfer during the shock-induced ignition of an explosive gas. International Journal of Heat and Mass Transfer, 267–276

Heywood JB, Higgins JM, Watts PA, Tabaczynski RJ (1979) Development and Use of a Cycle Simulation to Predict SI Engine Efficiency and NOx Emissions, SAE 790291 26 pp

Heywood, JB. (1988) Internal combustion engine fundamentals. McGraw-Hill Book Company, New York, xxix + 930 pp

Hindmarsh AL (1971) ODEPACK, A systematised collection of ODE solvers. Scientific Computing ed. R. F. Stephens, IMAC Trans. on Scientific Computing, North Holland, Amsterdam, pp 55–64

Hirschfelder JO, Curtis CF, Bird RB (1964) Molecular theory of gases and liquids. J. Wiley & Sons, New York, xxvi + 1249 pp

Hofbauer J (1956) The theory of evolution and dynamical systems: mathematical aspects of selection. Cambridge University Press, Cambridge, viii + 341 p

Horlock JH, Winterbone DE (ed) (1986) The thermodynamics and gas dynamics of internal combustion engines Clarendon Press, Oxford, vol.II, pp. 583–1237

Huang WM, Greif,R, Vosen SR (1987) The effects of pressure and temperature on heat transfer during flame quenching SAE, Paper 872106, 11 pp

Huang WM, Vosen,SR, Greif R (1987) Heat transfer during laminar flame quenching: effect of fuels. Twenty-First International Symposium on Combustion, The Combustion Institute, 1853–1860

Jante A (1960) The Wiebe Combustion Law (Das Wiebe-Brenngesetz, ein Forschritt in der Thermodynamik der Kreisprozesse von Verbrennungsmotoren) Kraftfahrzeugtchnik, v. 9, pp. 340–346

Jost, W (1946) Explosion and combustion processes in gases. McGraw-Hill Book Company, New York and London, xv + 621 pp

Kee R, Rupley FM, Miller JA (1993) The Chemkin thermodynamic data base. Sandia Report SAND87-8215

Kee RJ, Miller JA, Jefferson TH (1980) Chemkin: a general-purpose, problem-independent, transportable, Fortran chemical kinetics code package, Sandia Report, SAND80-8003

Kee RJ, Rupley FM, Miller JA (1989) CHEMKIN-II: A Fortran chemical kinetics package for the analysis of gas-phase chemical kinetics. Sandia Report SAND89-8009

Kolmogorov AN, Petrovskii, IG, Piskunov NS (1937) A Study of the Diffusion Equation with Increase in the Amount of Substance and Its Application to a Biological Problem Bull. Moscow Univ., Math. Mech., vol. 1, no. 6, pp.1–26,

[(1991) Selected Works of A.N. Kolmogorov, ed: Tikhomirov VM, Kluwer Academic Publishers, Dordrecht, v.1, Mathematics and Mechanics, pp. 242–270]

Kondrat'ev VN (1964) Chemical Kinetics of Gas Reactions (transl. J. M. Crabtree JM, Cation SN, ed. N. B. Slater) Pergamon Press, Oxford. Addison-Wesley Publishing, Reading, xiv + 812 pp

Krieger RB, Borman GL (1966) The computation of apparent heat release for internal combustion engines. ASME 66-WA/DGP-4, 16 pp

Lange W, Woschni G (1964) Thermodynamiosche Auswendung von Indikator-Diagrammen elektronisch gerechnet. MTZ 25: 284–289

Latsch R (1984) The Swirl-Chamber Spark Plug: Means of Faster More Uniform Energy Conversion in the Spark-Ignition Engine, SAE840455

Le Feuvre T, Myers PS, Uyehara OA (1969) Experimental instantaneous heat fluxes in a diesel engine and their correlation. SAE Paper 690464

Lefebvre A.H (1983) Gas Turbine Combustion. McGraw-Hill Co, vii + 531 pp

Lefebvre AH. (1989) Atomization and Sprays. Hemisphere Publishing Co., xi + 421 pp

Lewis, B, von Elbe, G (1987) Combustion, flames and explosion of gases, (esp. Chapter V, 15, Combustion Waves in Closed Vessels, pp. 381–395), Academic Press, Inc., Orlando, Florida (third edition), xxiv + 739 pp

List, H (1939, 1946) Thermodynamik der Verbrennungskraftmaschine, J. Springer. Wien, viii + 123 pp

Lotka AJ (1924) Elements of Mathematical Biology., Dover Publications reprint xxx + 465 pp

Lu JH, Ezekoye OA, Greif R, Sawyer RF (1991) Unsteady heat transfer during side wall quenching of a laminar flame. Twenty-Third International Symposium on Combustion, The Combustion Institute, 441–446

Maas U, Pope SB (1992a) Implementation of simplified chemical kinetics based on intrinsic low-dimensional manifolds. Twenty-Fourth Symposium (International) on Combustion, The Combustion Institute, Pittsburgh, PA, pp 103–112

Maas U, Pope SB (1992b) Simplifying chemical kinetics: intrinsic low-dimensional manifolds in composition space. Combustion and Flame, 88: 239–264

Maxson JA, Hensinger DM, Hom K, Oppenheim AK (1991) Performance of Multiple Stream Pulsed Jet Combustion Systems. SAE Paper 910565, 9 pp

McBride BJ, Gordon S (1996) Computer Program for Calculation of Complex Chemical Equilibrium Compositions and Applications, II. Users Manual and Program Description, NASA Reference Publication 1311, vi + 177 pp

Murase E, Ono S, Hanada K, Oppenheim AK (1994) Pulsed Combustion Jet Ignition in Lean Mixtures. SAE Paper 942048, 9pp

Murase E, Ono S, Hanada K, Oppenheim AK (1996) Initiation of Combustion by Flame Jets. Combustion Science and Technology, 113–114: 167–177

Najt PM, Foster DE (1983) Compression-Ignited Homogeneous Charge Combustion. SAE Paper 830264, 16 pp

Neumann K (1934) Influence of the Combustion Rate on the Work Process of a Compressor-less Diesel Engine (Einfluß der Verbrennungsgeschwindigkeit auf den Arbeitsproces eines kompresorlosen Dieselmotors") Forschung auf dem Gebiete des Ingenieurwesens 5

Noguchi M, Tanaka Y, Tanaka T, Takeuchi Y. (1979) A Study on Gasoline Engine Combustion by Observation of Intermediate Reactive Products during Combustion. SAE 790840, 13 pp

Obert EF. (1973) Internal combustion engines and air pollution Harper and Row, Publishers, New York, xiii + 740 pp. + 5 charts

Olofsson E, Alvestig P, Bergsten L, Ekenberg M, Gawell A, Larsén A, Reinman R (2001) A High Dilution Stoichiometric Combustion Concept Using a Wide Variable Spark Gap and in-Cylinder Air Injection in Order to Meet Future CO2 Requirements and World Wide Emission Regulations, SAE 2001-01-0246

Onishi S, Jo SH, Shoda K, Jo PD, Kato S (1979), Active Thermo-Atmospheric Combustion (ATAC) – A New Combustion Process for Internal Combustion Engines. SAE 790501, 10 pp

Oppenheim AK (`1992) The Future of Combustion in Engines. Proceedings of the Institution of Mechanical Engineers, C448/022, IMechE 1992-10: 187-192

Oppenheim AK (1984) The Knock Syndrome--Its Cures and Its Victims. SAE Paper 841339, 11 pp, SAE Transactions 93: 874–883

Oppenheim AK (1985) Dynamic features of combustion. Phil. Trans. Roy. Soc. London A 315: 471–508

Oppenheim AK (1988) Quest for Controlled Combustion Engines. SAE Transactions, The Journal of Engines 97: 1033–1039

Oppenheim AK (2002) Prospects for Combustion in Piston Engines SAE 2002-01-0999, 15pp

Oppenheim AK, Barton JE, Kuhl AL, Johnson WP (1997) Refinement of Heat Release Analysis. SAE 970538, 23 pp

Oppenheim AK, Beck NJ, Hom K, Maxson JA, Stewart HE (1990) A Methodology for Inhibiting the Formation of Pollutants in Diesel Engines. SAE 900394, 10 pp

Oppenheim AK, Beck NJ, Hom K, Maxson JA, Stewart HE (1990) A Methodology for Inhibiting the Formation of Pollutants in Diesel Engines. SAE 900394, 10 pp

Oppenheim A K, Beltramo J, Faris DW, Maxson JA, Hom K, Stewart HE (1989) Combustion by Pulsed Jet Plumes - Key to Controlled Combustion Engines. SAE 890153, SAE Transactions, The Journal of Engines 98: 175–182

Oppenheim AK, Cheng RK, Teichman K, Smith OI., Sawyer RF, Hom K., Stewart HE (1977) A Cinematographic Study of Combustion in an Enclosure Fitted with a Reciprocating Piston,.Stratified Charge Engines, IMechE Conference Publications 1976-11, The Institution of Mechanical Engineers, London, pp. 127–135

Oppenheim AK, Kuhl AL (1995) Paving the Way to Controlled Combustion Engines (CCE). SAE 951961, Futuristic Concepts in Engines and Components, SAE SP-1108: 19–29

Oppenheim AK, Kuhl AL (1998) Life of Fuel in Engine Cylinder. SAE 980780, Modeling of SI and Diesel Engines. SAE SP-1330: 75-84; SAE Transactions, Journal of Engines 103: 1080–1089

Oppenheim AK, Kuhl AL (2000a) Energy Loss from Closed Combustion Systems. Proceedings of the Combustion Institute 28:1257-1263

Oppenheim AK, Kuhl AL (2000b) Dynamic Features of Closed Combustion Systems. Progress in Energy and Combustion Science 26: 533–564

Oppenheim AK, Kuhl AL, Packard A.K., Hedrick JK, Johnson WP (1996) Model and Control of Heat Release in Engines SAE 960601, Engine Combustion and Flow Diagnostics SAE SP-1157: 15–23

Oppenheim AK, Maxson JA (1994) A thermochemical phase space for combustion in engines. Twenty-Fifth Symposium (International) on Combustion, The Combustion Institute, Pittsburgh, Pennsylvania, pp. 157–165

Oppenheim, A.K., Maxson, J.A. and Shahed S.M (1994) Can the Maximization of Fuel Economy be Compatible with the Minimization of Pollutant Emissions? SAE 940479, 12 pp

Oppenheim AK, Spektor R, Sum T-HJ, Kuhl AL, Johnson WP (2000) Dynamics of Combustion in a Diesel Engine Under the Influence of Air/Fuel Ratio. SAE 2000-01-0203, SAE SP-1492: 135–148

Oppenheim AK, Sum T-HJ, Gebert K, Johnson WP, Kuhl AL (2001) Influence of Charge Dilution on the Dynamic Stage of Combustion in a Diesel Engine SAE 2001-01-0551, Advances in Combustion 2001 SP. 1574: 181–188

Oppenheim AK, Teichman K, Hom K, Stewart HE (1978) Jet Ignition of an Ultra-Lean Mixture, SAE 780637, 13 pp

Peters N (2000) Turbulent combustion, Cambridge University Press, xvi + 304 pp

Pischinger A, Cordier 0 (1939) Gemischbildung und Verbrennung im Dieselmotor, Springer, Wien, viii + 128 pp

Pischinger R., Klell M, Sams T (1989–2002) Thermodynamik der Verbrennungs-Kraftmaschine, 2nd edition, Der Fahrzeugantrieb. Springer, Wien NewYork, xvii+475 pp

Pischinger, A (1948) Die Steuerung der Verbrennungskraft- maschinen, J. Springer, Wien, vii + 239pp

Poincaré H (1892) Thermodynamique, Gothiers—Villars, Paris, xix + 432 pp. [1908 edition, xix + 458 pp]

Rashevsky N.(1948) Mathematical Biophysics The University of Chicago Press, xxiii+669 pp

Rassweiler GM, Withrow L (1938) Motion Pictures of Engine Flames Correlated with Pressure Cards. SAE 800131: 20 pp

Reynolds WC (1996) STANJAN interactive computer programs for chemical equilibrium analysis, Department of Mechanical Engineering, Stanford University, Stanford, California, 48 pp

Ricardo HR (1922–1923) The high-speed internal-combustion engine. Blackie & Son, Ltd., London

Saas F (1929) Kompressorlose Dieselmaschinen (Druckeinspritzmaschinen): ein Lehrbuch für Studierende. J. Springer, Berlin, 395 pp

Schmidt FAF (1951) Verbrennungskraftmaschinen: Thermodynamik und versuchsmäßige Grundlagen der Verbrennungsmotoren und Gasturbinen. R. Oldenbourg, Munich, 427 pp. [transl. Mitchell RWS, Horne J. (1965) The internal combustion engine. Chapman and Hall, London, xxiii + 579 pp]

Sell GR (1937) Dynamics of evolutionary equations. Springer, New York, xiii + 670 pp

Semenoff NN (1934) Chain Reactions Goskhimtekhizdat, Leningrad, [transl.: Chemical Kinetics and Chain Reactions (1935) Oxford University Press

Semenov NN (1958-59) Some Problems in Chemical Kinetics and Reactivity" Princeton University Press, Princeton,. 1: xii+239 pp; 2: v+331 pp

Shen Y, Schock H.J, Oppenheim AK (2003) Pressure Diagnostics of Closed System in a Direct Injection Spark Ignition Engine. SAE 2003-01-0723, 11pp

Shen Y, Schock HJ, Sum T-HJ, Oppenheim AK (2002) Dynamic Stage of Combustion in a Direct Injection Methanol Fueled Engine. SAE 2002-01-0998, 18 pp

Sirignano WA (1999) Fluid Dynamics and Transport of Droplets and Sprays. Cambridge University Press, xvii + 311 pp

Smy PR, Clements, RM, Oppenheim AK, Topham DR (1997) Structure of the Pulsed Plasma Jet. J. Phys. D: Appl. Phys. 20: 1016–1020

Spalding B (1957) I. Predicting the Laminar Flame Speed in Gases with Temperature-explicit Reaction Rates. Combustion and Flame 1: 287–295; II. One-dimensional Laminar Flame Theory for Temperature-explicit Reaction Rates" Combustion and Flame, 1: 296–307

Stan C (1999) Direkteinspritzsysteme fur Otto- und Dieselmotoren. Springer-Verlag, Berlin Heidelberg [transl. (2000) Direct Injection Systems for Spark-Ignition and Compression-Ignition Engines SAE, Inc. Warrendale, x + 288]

Stan C (2003) Direct Injection Systems for Internal Combustion Engines SAE, Inc. Warrendale

Steinfeld JI, Francisco JS, Hase WL (1989) Chemical Kinetics and Dynamic. Prentice Hall, New Jersey x + 518 pp

Stull DR, Prophet H (1971) JANAF Thermochemical tables. National Bureau of Standards (currently National Institute of Standards and Technology), US Department of Commerce) Report NSRDS-NBS 37, 1141 pp

Taylor CF. (1982-1985) The internal combustion engine in theory and practice, 1 Thermodynamics, Fluid Flow, Performance. (2nd edition, x + 574 pp.; 2 (5th printing, 1982): Combustion, Fuels, Materials, Design, MIT Press, Cambridge, Massachusetts x + 783 pp

Vibe II (1956) Semi-Empirical Expression for Combustion Rate in Engines. Proceedings of Conference on Piston Engines, USSR Academy of Sciences, Moscow, pp. 185-191.

Vibe II (1970) Progress of Combustion and Cycle Process in Combustion Engines .transl. Heinrich J Brennverlauf und Kreisprozeß von Verbrennungsmotoren VEB Verlag Technik, Berlin, 286 pp.

Volterra V (1937) Principles of Biological Mathematics:Principes de Biologie Mathematique: Premier Partie" Acta Biotheoretica, v. 3, no. 1

Vosen SR, Greif R, Westbrook CK (1985) Unsteady heat transfer during laminar flame quenching. Twentieth International Symposium on Combustion, The Combustion Institute, 75–83

Warnatz J, Maas U, Dibble RW (1996) Combustion. Physical and chemical fundamentals, modeling and simulation, experiments, pollutant formation. Springer, Berlin, x + 265 pp

Westbrook C K, Pitz WJ (1984) A comprehensive chemical kinetic reaction mechanism for oxidation and pyrolysis of propane and propene. Combustion Science and Technology, 37: 117–152

Williams FA (1985) Combustion Theory. The Benjamin/Cummings Publishing Company, Menlo Park, California (second edition) xxiii + 680 pp

Withrow L, Rassweiler GM (1936) Slow motion shows knocking and non-knocking explosions. *SAE Transactions* 39: 297–303, 312

Woschni G (1965a) Beitrag zum Problem des Wärmeüberganges im Verbrennungsmotor MTZ 26, 128–133

Woschni G (1965b) Elektronische Brechnung von Verbrennungsmotor-Kreisprozessen. MTZ 26: 439–446

Woschni G (1966/7) Experimentelle Untersuchungen zum instatonären Wärmeübergang während der Verbrennung bei konstanten Volumen, M.A.N. Forschungsheft 13

Woschni G (1967) Universally applicable equation for the instantaneous heat transfer coefficient in the internal combustion engine. SAE Paper 670931, SAE Trans. 76

Woschni G (1970) Die Berechnung der Wandverluste und der thermischen Belastung der Bauteile von Dieselmotoren. MTZ 31: 491–499

Woschni G (1987) Engine cycle simulation - an effective tool for the development of medium speed diesel engines. SAE Paper 870570, SAE Trans. 96

Woschni G, Anisits F (1973) Eine Methode zur Vorausberechnung des Brennverlaufs mittelschnellaufender Dieselmotoren bei geänderten Betriebsbedingungen. MTZ 34: 106–115

Woschni G, Anisits F (1974) Experimental investigation and mathematical representation of the rate of heat release in diesel engines dependent on engine operating conditions. SAE Paper 740086

Woschni G, Fieger F (1979) Determination of local heat transfer coefficients at the piston of a high speed Diesel engine by evaluation of measured temperature distribution. SAE Paper 790834, SAE Trans. 88

Woschni G, Kolesa K, Spindler W (1986) Isolierung der Brennraumme-wände - Ein lohnendes Entwicklungsziel bei Verbrennungsmotoren. MTZ 47: 495-500

Yang CH, Gray BF (1967). The determination of explosion limits from a unified thermal chain theory. Eleventh Symposium (International) on Combustion, 1099-1106, The Combustion Institute, Pittsburgh

Zel'dovich YaB (1941) On the Theory of Thermal Intensity: Exothermic Reactions in a Jet. Zhurnal Techniskoi Fiziki 11: 493-500

Zel'dovich YaB, Barenblatt GI., Librovich VB, Makhviladze GM (1980) Matematicheskaya teoriya goreniya i vzryva. Nauka, USSR Academy of Sciences, Moscow, 478 pp [trasnsl. McNeil DH (1985) The mathematical theory of combustion and explosions. (vid. esp. Chapter 6, Combustion in Closed Vessels. pp. 470–487) Consultants Bureau, New York and London, xxi + 597 pp]

Zel'dovich YaB, Frank-Kamenetskii DA (1938) Teoriya teplovogo rasprotraneniya plameni (A Theory of Thermal Flame Propagation) Zhurnal Fizicheskoikhimii, vol. 12, no.1, pp.100–105

Nomenclature

Symbols

$C_K \equiv (\partial e_K / \partial w_K)_p$ – gradient of a state vector

$c_{Kp} \equiv (\partial h_K / \partial T)_p$ – specific heat at constant volume

$c_{Kv} \equiv (\partial e_K / \partial T)_v$ – specific heart at constant pressure

e_K internal energy

h_K enthalpy

n_k polytropic index

M_K molecular mass

$m_k \equiv 1 - n_k^{-1}$

p pressure

$P \equiv p/p_i$ – normalized pressure

$q_R \equiv u_{Ro} - u_{Po}$ - reference exothermic energy

$R_K \equiv R/M_K$

R gas constant

t time

T_K temperature

$\tilde{T}_K \equiv T_K / T_i$

u_K internal energy in thermodynamic tables

v_K specific volume

$\tilde{v}_K \equiv v_K / v_{Ki}$

$w_K \equiv p_K v_K$ – mechanical energy

$W_K \equiv w_K / w_{Si}$

x progress parameter

x_π progress parameter of π

y_L mass fraction of L or its progress parameter

$\tilde{y}_L \equiv y_L / y_{Lmax}$

Y_R mass fraction of consumed reactants or generated products

z_K w_K, u_K

α life function coefficient

$\beta \equiv \chi + 1$ index of Vibe function

χ index of life function

$\gamma_K \equiv c_{Kp}/c_{Kv}$ specific heats ratio
$\lambda \equiv \sigma/\sigma_{\text{stoichiometric}}$ – air excess factor
ν stoichiometric coefficient
π polytropic pressure model
θ crank angle
ρ density
σ air/fuel mass ratio
τ time normalized with respect to lifetime
ζ life function exponent

Designations

A   air
B   inert component
C   charge
c   compression
e   expansion
E   effective
F   fuel
f   final
i   initial
I   ineffective
K   A, F, R, B, C, P
L   P, E, I
R   reactants
P   products
s   surroundings
S   system
t   terminal

Index

Printing: Mercedes-Druck, Berlin
Binding: Stein+Lehmann, Berlin